AF508668

Storm Tide and Wave Simulations and Assessment

Storm Tide and Wave Simulations and Assessment

Editors

Shih-Chun Hsiao
Wen-Son Chiang
Wei-Bo Chen

MDPI • Basel • Beijing • Wuhan • Barcelona • Belgrade • Manchester • Tokyo • Cluj • Tianjin

Editors
Shih-Chun Hsiao
National Cheng Kung University
Taiwan

Wen-Son Chiang
National Cheng Kung University
Taiwan

Wei-Bo Chen
National Science and Technology Center for Disaster Reduction
Taiwan

Editorial Office
MDPI
St. Alban-Anlage 66
4052 Basel, Switzerland

This is a reprint of articles from the Special Issue published online in the open access journal *Journal of Marine Science and Engineering* (ISSN 2077-1312) (available at: https://www.mdpi.com/journal/jmse/special_issues/zara_storm_tide_simulations_assessment).

For citation purposes, cite each article independently as indicated on the article page online and as indicated below:

LastName, A.A.; LastName, B.B.; LastName, C.C. Article Title. *Journal Name* **Year**, *Volume Number*, Page Range.

ISBN 978-3-0365-0496-4 (Hbk)
ISBN 978-3-0365-0497-1 (PDF)

Contents

About the Editors

Shih-Chun Hsiao is a professor and currently the chair of the Department of Hydraulic and Ocean Engineering, National Cheng Kung University (Taiwan). He completed his Ph.D. at the Department of Civil and Environmental Engineering, Cornell University in 2000. His research interests are coastal engineering, porous medium flow, coastal risk and management, and the modeling of storm surges and waves.

Wen-Son Chiang is a research fellow and the Deputy Director of Tainan Hydraulics Laboratory, National Cheng Kung University. He has a large amount of experience in hydraulic laboratory tests for wave properties. His research interests are nonlinear wave dynamics, coastal erosion, marine sediment, and submarine landslides.

Wei-Bo Chen is a research fellow of National Science and Technology Center for Disaster Reduction, Taiwan. He completed his Ph.D. at the Department of Bioenvironmental Systems Engineering, National Taiwan University in 2012. His research interests are estuarine dynamics modeling, city-scale inundation modeling, storm surge and wave modeling, wave climate, marine renewable energy analysis, and marine weather forecasting.

Preface to "Storm Tide and Wave Simulations and Assessment"

Storm tides, surges, and waves associated with typhoons/tropical cyclones/hurricanes are among the most severe threats to coastal zones, nearshore waters, and navigational safety. Therefore, predicting typhoon/tropical cyclone/hurricane-induced storm tides, surges, waves, and coastal erosion is important to reduce loss of human life and property and to mitigate coastal disasters. Although many studies on hindcasting/predicting/forecasting of typhoon-driven storm tides, surges, and waves as well as the morphological evolution have been carried out through numerical models in the last decade, there is still a growing demand for novel techniques which could be adopted to resolve the complex physical processes of storm tides, surges, waves, and coastal erosion. To improve and enhance our simulating and analytic capabilities and understanding of storm tides, surges, and waves, this Special Issue is intended to collect the latest studies or reviews on storm tides, surges, and waves modeling and analysis utilizing dynamic and statistical models and artificial intelligence approaches. Seven high-quality papers have been published in this Special Issue which cover the application and development of many high-end techniques for storm tides, surges, and waves.

Shih-Chun Hsiao, Wen-Son Chiang, Wei-Bo Chen
Editors

Journal of
Marine Science and Engineering

Editorial

Storm Tide and Wave Simulations and Assessment

Shih-Chun Hsiao [1], Wen-Son Chiang [2] and Wei-Bo Chen [3,*

[1]   Department of Hydraulic and Ocean Engineering, National Cheng Kung University, Tainan 701, Taiwan; schsiao@mail.ncku.edu.tw
[2]   Tainan Hydraulics Laboratory, National Cheng Kung University, Tainan 70101, Taiwan; chws@mail.ncku.edu.tw
[3]   National Science and Technology Center for Disaster Reduction, New Taipei 231, Taiwan
*   Correspondence: wbchen@ncdr.nat.gov.tw; Tel.: +886-2-8195-8612

Received: 12 January 2021; Accepted: 12 January 2021; Published: 14 January 2021

Keywords: numerical modeling; statistical analysis; artificial intelligence techniques; storm tide; storm surge; storm wave; coastal morphology

1. Introduction

Storm tides, surges, and waves associated with typhoons/tropical cyclones/hurricanes are among the most severe threats to coastal zones, nearshore waters, and navigational safety. Therefore, predicting typhoon/tropical cyclone/hurricane-induced storm tides, surges, waves, and coastal erosion is important to reduce loss of human life and property and to mitigate coastal disasters. Although many studies on hindcasting/predicting/ forecasting of typhoon-driven storm tides, surges, and waves as well as the morphological evolution have been carried out through numerical models in the last decade, there is still a growing demand for novel techniques which could be adopted to resolve the complex physical processes of storm tides, surges, waves, and coastal erosion.

To improve and enhance our simulating and analytic capabilities and understanding of storm tides, surges, and waves, this Special Issue is intended to collect the latest studies or reviews on storm tides, surges, and waves modeling and analysis utilizing dynamic and statistical models and artificial intelligence approaches. Seven high-quality papers have been published in this Special Issue which cover the application and development of many high-end techniques for storm tides, surges, and waves: for instance, employment of an artificial neural network for predicting coastal freak waves [1]; reproduction of super typhoon-created extreme waves [2]; numerical analysis of nonlinear interactions for storm waves, tides, and currents [3]; wave simulation for an island using a circulation-wave coupled model [4]; analysis of typhoon-induced waves along typhoon tracks in the western North Pacific Ocean [5]; understanding of how a storm surge prevents or severely restricts aeolian supply [6]; and investigation of coastal settlements and assessment of their vulnerability [7].

2. Details of the Papers

Doong et al. [1] propose a probabilistic coastal freak wave (CFW) forecasting model that is an advancement of a previously proposed deterministic CFW forecasting model. They also develop a probabilistic forecasting scheme to make an artificial neural network (ANN) model achieve probabilistic CFW forecasting.

Hsiao et al. [2] employ the Climate Forecast System Version 2 (CFSV2) winds under varying spatial and temporal resolutions to simulate large waves driven by a duper typhoon. They indicate that the simulated storm wave heights tend to decrease significantly due to the lower spatial resolution of hourly winds from the CFSV2 dataset; however, the variations in the storm wave height simulations were less sensitive to the temporal resolution of the wind field.

Hsiao et al. [3] use a high-resolution, unstructured-grid, coupled circulation-wave model (Semi-implicit Cross-scale Hydroscience Integrated System Model Wind Wave Model version III (SCHISM-WWM-III)) to investigate the sensitivity of storm wave simulations to storm tides and tidal currents. They find that the simulated significant wave height, mean wave period, and wave direction for a wave buoy in the outer region of a typhoon are more sensitive to tidal current but less sensitive to tidal elevation than those for a wave buoy moored in the inner region of a typhoon and also suggest that the inclusion of tidal current and elevation could be more important for typhoon wave modeling in sea areas with larger tidal ranges and higher tidal currents.

Yang et al. [4] apply a numerical wave model Simulating Waves Nearshore (SWAN), which resolves nearshore wave processes, and a hydrodynamic model, the Finite-Volume Community Ocean Model (FVCOM) to simulate waves and currents during Typhoon Fung-wong (2014) and Typhoon Chan-hom (2015) around the Zhoushan Islands. The influence of sea-surface currents, e.g., typhoon-induced and tidal currents, as well as the sea-water level, on wave simulation was also studied.

Hu et al. [5] use version 5.16 of the WAVEWATCH-III (WW3) model to simulate parameters of typhoon-generated wave fields in the western North Pacific Ocean during the period 1998–2017. Overall, a typhoon-induced wave energy dominates north of 30° N. Temporal analysis of the leading principal component of significant wave height indicates that (a) the intensity of the wave pattern produced by westward-tracking typhoons decreased during the last 20 years and (b) typhoons that recurve east of 140° E and those that track westward toward southeast Asia are largely responsible for the decadal variability of typhoon-induced wave distribution.

Tuijnman et al. [6] predict aeolian supply during a 2-day surge period to be about 66% of the potential supply using a fetch-based model. Fetch limitations imposed by the surge-induced inundation and the continuous saturation of sand on the emerging beach both contributed to the predicted supply limitation. Their results quantitatively support earlier studies that suggested surges to be the primary condition that causes predictions of long-term potential foredune growth to overestimate measured growth.

Lin et al. [7] use the east coast of Taiwan as an example, through geographic information system (GIS) and statistical analysis in land-use status, vulnerable population groups and Unmanned Aerial Vehicle (UAV) landscape signs of indicators of erosion and accumulation. Through the main output of an intuition scatter map, the erosion landscape susceptibility, economical land-use exposure, and special population groups' ratio allowed for easy comparison of the vulnerability, risk level, and resilience between different coastal settlements.

Funding: This research received no external funding.

Conflicts of Interest: The authors declare no conflict of interest.

References

1. Dong, J.D.; Chen, S.T.; Chen, Y.C.; Tsai, C.H. Operational Probabilistic Forecasting of Coastal Freak Waves by Using an Artificial Neural Network. *J. Mar. Sci. Eng.* **2020**, *8*, 165. [CrossRef]
2. Hsiao, S.C.; Chen, H.; Wu, H.L.; Chen, W.B.; Chang, C.H.; Guo, W.D.; Chen, Y.M.; Lin, L.Y. Numerical Simulation of Large Wave Heights from Super Typhoon Nepartak (2016) in the Eastern Waters of Taiwan. *J. Mar. Sci. Eng.* **2020**, *8*, 217. [CrossRef]
3. Hsiao, S.C.; Wu, H.L.; Chen, W.B.; Chang, C.H.; Lin, L.Y. On the Sensitivity of Typhoon Wave Simulations to Tidal Elevation and Current. *J. Mar. Sci. Eng.* **2020**, *8*, 713. [CrossRef]
4. Yang, Z.; Shao, W.; Ding, Y.; Shi, J.; Ji, Q. Wave Simulation by the SWAN Model and FVCOM Considering the Sea-Water Level around the Zhoushan Islands. *J. Mar. Sci. Eng.* **2020**, *8*, 783. [CrossRef]

5. Hu, Y.; Shao, W.; Wei, Y.; Zuo, J. Analysis of Typhoon-Induced Waves along Typhoon Tracks in the Western North Pacific Ocean, 1998–2017. *J. Mar. Sci. Eng.* **2020**, *8*, 783. [CrossRef]
6. Tuijnman, J.T.; Donker, J.J.A.; Schwarz, C.S.; Ruessink, G. Consequences of a Storm Surge for Aeolian Sand Transport on a Low-Gradient Beach. *J. Mar. Sci. Eng.* **2020**, *8*, 584. [CrossRef]
7. Lin, S.W.; Yen, C.F.; Chang, C.H.; Wang, L.J.; Shih, H.J. Comprehensive Natural Environment and Landscape Signs in Coastal Settlement Hazard Assessment: Case of East Taiwan between the Coastal Mountain and the Pacific Ocean. *J. Mar. Sci. Eng.* **2020**, *8*, 478. [CrossRef]

Article

Operational Probabilistic Forecasting of Coastal Freak Waves by Using an Artificial Neural Network

Dong-Jiing Doong [1], Shien-Tsung Chen [1,*], Ying-Chih Chen [1] and Cheng-Han Tsai [2]

[1] Department of Hydraulic and Ocean Engineering, National Cheng Kung University, Tainan City 701, Taiwan; doong@mail.ncku.edu.tw (D.-J.D.); venwhah@gmail.com (Y.-C.C.)
[2] Department of Marine Environmental Informatics, National Taiwan Ocean University, Keelung City 202, Taiwan; chtsai@mail.ntou.edu.tw
* Correspondence: chen@gs.ncku.edu.tw

Received: 8 February 2020; Accepted: 27 February 2020; Published: 3 March 2020

Abstract: Coastal freak waves (CFWs) are unpredictable large waves that occur suddenly in coastal areas and have been reported to cause casualties worldwide. CFW forecasting is difficult because the complex mechanisms that cause CFWs are not well understood. This study proposes a probabilistic CFW forecasting model that is an advance on the basis of a previously proposed deterministic CFW forecasting model. This study also develops a probabilistic forecasting scheme to make an artificial neural network model achieve the probabilistic CFW forecasting. Eight wave and meteorological variables that are physically related to CFW occurrence were used as the inputs for the artificial neural network model. Two forecasting models were developed for these inputs. Model I adopted buoy observations, whereas Model II used wave model simulation data. CFW accidents in the coastal areas of northeast Taiwan were used to calibrate and validate the model. The probabilistic CFW forecasting model can perform predictions every 6 h with lead times of 12 and 24 h. The validation results demonstrated that Model I outperformed Model II regarding accuracy and recall. In 2018, the developed CFW forecasting models were investigated in operational mode in the Operational Forecast System of the Taiwan Central Weather Bureau. Comparing the probabilistic forecasting results with swell information and actual CFW occurrences demonstrated the effectiveness of the proposed probabilistic CFW forecasting model.

Keywords: coastal freak wave; probabilistic forecasting; artificial neural network

---

1. Introduction

A coastal freak wave (CFW) is a sudden, unpredictable large wave that occurs in coastal areas as a result of complex interactions between shoaling waves and coastal topography. CFWs can occur even when the sea is calm and in areas without previous notable CFWs. Although CFWs are uncommon, they have been reported in a variety of areas [1–6]. CFWs can cause numerous casualties when people are present in the area where the CFW occurs. Therefore, CFW forecasting and timely warning information can mitigate the potential harm of CFWs. Numerical models based on the wind wave energy spectrum are usually applied for forecasting wave characteristics. However, the complex mechanisms that cause CFWs may include the propagation of nearshore waves, wave structure interactions, and topographic bathymetry effects, rendering CFWs difficult to simulate using physical laws. Therefore, numerical models are currently unable to simulate and forecast CFWs. Nevertheless, some recent works on the characterization of large scale extreme events in geophysical flows shed a new light on the CFW genesis and forecasting [7]. Simulating and forecasting CFWs in a deterministically physical method may be attainable in the near future.

Artificial intelligence (AI) methods can be used for CFW simulation and forecasting. Various AI methods have been applied to hydrologic and ocean forecasting. In hydrologic forecasting, artificial neural networks [8–12], support vector machines [13–20], and fuzzy set theory [21–27] have been successfully used to perform rainfall and flood forecasting. In ocean forecasting, AI methods have been employed to forecast significant wave height. For example, Deo and Sridhar Naidu [28], Deo et al. [29], Tsai et al. [30], and Mandal and Prabaharan [31] applied artificial neural networks to predict wave height on the basis of wave height data and observed or simulated wind speed. Mahjoobi and Mosabbeb [32] and Elbisy [33] employed support vector machines to simulate significant wave height and wave characteristics. Moreover, Kazeminezhad et al. [34], Özger and Şen [35], and Mahjoobi et al. [36] used fuzzy logic models to simulate or forecast wave height, and Chang and Chien [37,38], Tsai et al. [39], Wei [40], and Chen [41] applied typhoon characteristics as the inputs of AI models to forecast wave parameters during typhoons.

However, few studies have applied AI methods to CFW forecasting. Doong et al. [6] first proposed applying artificial neural networks to CFW forecasting. Their proposed model adopted various sea state characteristics to forecast CFW occurrence. The methodology is similar to a clustering process that deterministically predicts whether a CFW will or will not occur. The present study improved on the deterministic methodology used by Doong et al. [6] by developing a probabilistic CFW forecasting model, which is an advance to fulfill the probabilistic forecasting of CFW occurrence. This probabilistic forecasting model is more practical than the deterministic model, and agencies can use it to establish risk-based criteria for warnings and responses.

A back-propagation neural network (BPNN) was used to develop the probabilistic CFW forecasting model. This study proposed a novel probabilistic scheme that assigns the BPNN outputs to probability values indicating the physical tendency of a CFW to occur. Variables potentially related to CFW occurrence were selected as the model inputs on the basis of current knowledge of CFWs. The model output is a numerical value ranging from 0 to 1 that indicates the probability of CFW occurrence. Probabilistic CFW forecasting was conducted using validation data, and an operational system was developed and tested to examine the effectiveness of the developed CFW forecasting model. Validation results based on real CFW data confirmed the predictive ability of the proposed probabilistic CFW forecasting model. The CFW forecasting model was tested in operational mode in the Operational Forecast System of the Taiwan Central Weather Bureau (TCWB) during 2018. Comparing the forecasting results with swell information and actual CFWs during a typhoon period demonstrated the effectiveness of the proposed probabilistic CFW forecasting model.

2. Wave Data and Probabilistic Forecasting

2.1. CFW Data

Coastal fishing is a popular activity in Taiwan, and people fishing in coastal areas sometimes encounter CFWs, which have occasionally resulted in casualties. CFW accidents in Taiwan occur the most frequently in the Longdong headland of northeast Taiwan (Figure 1). During 2000–2016, 63 CFW accidents were reported in the Longdong headland and its neighboring coast. These CFW accidents, which were checked and validated by the TCWB, were used as CFW data in this study. A CFW accident refers to a CFW that results in casualties. When a CFW does not result in injury, it is not recorded as a CFW accident. Accordingly, the number of CFW events is considerably higher than that of CFW accidents.

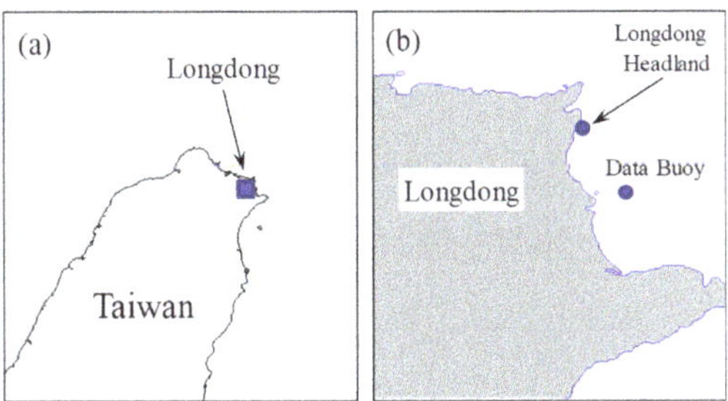

Figure 1. Map of (**a**) Longdong headland and (**b**) the Taiwan Central Weather Bureau (TCWB) data buoy.

2.2. Buoy Observation and Wave Model Simulation Data

The TCWB operates a data buoy in the Longdong area to record meteorological and wave data at intervals of 1 h. The buoy is located at 121.92° E and 25.10° N, which is approximately 1 km from the coast and has a water depth of approximately 23 m (Figure 1). This data buoy measures meteorological variables, including wind speed, wind direction, air pressure, air temperature, and surface water temperature, and wave parameters, such as the wave height, period, and direction. A data quality control procedure proposed by Doong et al. [42] was adopted to ensure the quality of the hourly observations of meteorological and wave data.

In addition to data buoy observations, the TCWB uses the WaveWatch III (WWIII) model to simulate the marine meteorology in the western Pacific. The spatial resolution of TCWB's WWIII model is 25 km for coastal ocean around Taiwan. A directional wave spectrum at each grid is available at hourly time steps. This study used an interpolation method to derive the directional wave spectrum at the buoy location. Therefore, the meteorological and wave variables from the WWIII model are also available for CFW prediction. The hourly meteorological and wave data from the data buoy and wave variables from the WWIII model were collected and organized for a period covering the studied CFW accidents.

2.3. Back-Propagation Neural Network

This study adopted a BPNN, which is the most commonly used type of artificial neural network. BPNNs feature a feedforward network structure with input, hidden, and output layers. BPNNs are trained using the back-propagation algorithm—a supervised learning method based on the method of steepest descent [43]. The model is briefly introduced as follows, and detailed descriptions of the model and its development are available in previous studies [44,45]:

Let X_i ($i = 1, 2, \ldots, m$) be the input variables in the input layer and $\hat{Y}_k$ ($k = 1, 2, \ldots, p$) be the output variables in the output layer. A BPNN with n neurons in the hidden layer is formulated as follows:

$$\hat{Y}_k = \sum_{j=1}^{n} W_{jk} \cdot f\left(\sum_{i=1}^{m} W_{ij} \cdot X_i + b_j\right) + d_k \tag{1}$$

where W_{ij} is the weight of the connection between the ith input neuron and the jth hidden neuron ($j = 1, 2, \ldots, n$), b_j is the bias term of the jth hidden neuron, W_{jk} is the weight of the connection between the jth hidden neuron to the kth output neuron, d_k is the bias term of the kth output neuron, and f is the activation function of the hidden neurons. In the present study, the hyperbolic tangent sigmoid transfer function was adopted as the activation function.

$$f(x) = \frac{1 - e^{-2x}}{1 + e^{-2x}} \tag{2}$$

The output of the BPNN $\hat{Y}_k$ is fit to the actual output Y_k by minimizing the objective function E:

$$E = \frac{1}{2}\sum_{k=1}^{p}\left(\hat{Y}_k - Y_k\right)^2 \tag{3}$$

During model training, the method of steepest descent is used to adjust the weights and biases of the network. Adjustments are repeated until an acceptable convergence of the weights and biases is reached.

2.4. Probabilistic Forecasting

Probability is the measure of the likelihood of an event. In this study, probability is regarded as the physical tendency of a CFW to occur. Probability values range from 0 to 1, where 0 implies impossibility and 1 implies certainty. The higher the probability of a CFW occurrence, the more likely it is that the CFW event will occur. Figure 2 illustrates the process of the probabilistic forecasting method used in this study. On the basis of the input of relevant variables (selection detailed in Section 3.1), the BPNN model outputs a probability value for the occurrence of a CFW. A high probability value (e.g., 0.91) indicates that a CFW is highly likely to occur. A low probability value (e.g., 0.18) indicates that a CFW is highly unlikely to occur.

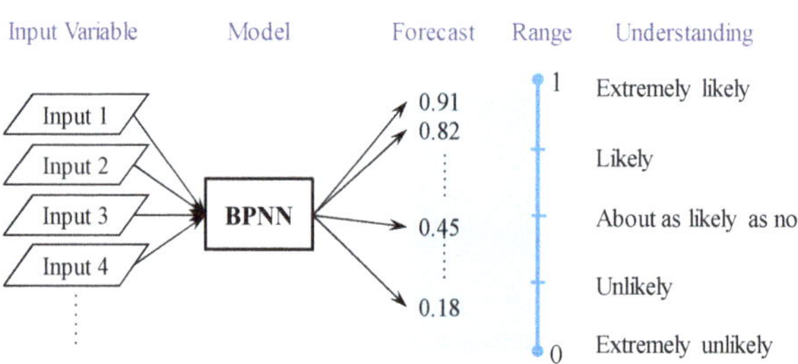

Figure 2. Probabilistic forecasting method.

3. Model Development

3.1. Selection of Input Variables

A probabilistic CFW forecasting model was developed to forecast the likelihood of CFW occurrence in the subsequent time interval. Input variables should be physically related to CFW occurrence. Eight meteorological and wave variables, namely significant wave height, peak wave period, onshore wind speed, misalignment between wind and wave directions, kurtosis of sea surface elevation, Benjamin–Feir index (BFI), groupiness factor, and abnormality index, were selected on the basis of expert knowledge and a preliminary experiment. These variables were used in the BPNN model to predict the probability of CFW occurrence. With the exception of misalignment between wind and wave directions, maximum values during the 6 h period before CFW occurrence were used as the model input. For misalignment between wind and wave directions, the value during CFW occurrence was used. The variables are detailed as follows.

3.1.1. Significant Wave Height

Severe sea states feature large waves and an increased probability of CFW occurrence. Significant wave height is the basic variable describing sea states and is defined as the mean of the highest third wave heights of all waves during a certain period.

3.1.2. Peak Wave Period

Chien et al. [1] and Tamura et al. [46] identified a strong relationship between swell and CFW occurrence. Swell has a period longer than that of waves caused by wind. It can transfer more energy and has more momentum compared with wind waves, and it can create a large spray of water when it breaks at the coast. Swell is represented by the peak wave period, which is equivalent to the period with the highest energy.

3.1.3. Onshore Wind Speed

Strong wind can generate large wind waves in coastal areas, especially when the wind direction is perpendicular to the coast. When the wind speed is high, large wind waves can increase the probability of CFW occurrence.

3.1.4. Misalignment between Wind and Wave Directions

As mentioned in the previous variable descriptions, wind waves and swell may produce CFWs. Gramstad and Trulsen [47] indicated that the combination of local wind waves and swell can increase the probability of CFWs by 5%–20%. When wind waves and swell propagate in similar directions, the probability of a CFW is higher. Therefore, the difference between local wind and swell directions is adopted as an input variable.

3.1.5. Kurtosis of Sea Surface Elevation

A CFW is an extreme outlier in a wave time series. Kurtosis measures the type of tail of a probability distribution. Mori et al. [48] discussed the relationship between the probability of CFWs and kurtosis, and they indicated that wave kurtosis is a suitable input for CFW forecasting.

3.1.6. Benjamin–Feir Index

When wave nonlinearity and instability increase, wave coalescence can produce CFWs. The BFI [49] is the ratio of wave steepness to spectral bandwidth, which quantitatively measures wave instability and nonlinearity. A higher BFI corresponds to a higher probability of CFW occurrence. Janssen [50] and Mori and Janssen [51] have provided details of BFI computation.

3.1.7. Groupiness Factor

Chien et al. [1] and Tsai et al. [2] claimed that a strong connection exists between wave groupiness and CFW occurrence. When a wave breaks at the coast, interaction with subsequent waves in the wave group may produce a CFW. The groupiness factor, proposed by Funke and Mansard [52], is adopted to quantify wave grouping. Sea surface elevation data obtained from buoy observations can be used to calculate the groupiness factor. However, the groupiness factor cannot be computed from the WWIII model outputs that do not include sea surface elevation data. For these outputs, the present study employed the peakedness factor [53], which is related to wave energy and groupiness. Wave energy is higher when the wave grouping phenomenon is more prominent. Thus, the peakedness factor replaced the groupiness factor when WWIII model data were used.

3.1.8. Abnormality Index

When sea states include abnormal waves, CFWs are likely to occur. The abnormality index [54,55] is the ratio of the wave height to the significant wave height. An extreme wave event is defined as a wave event with an abnormality index higher than 2. The abnormality index can be calculated from buoy observations but cannot be obtained from WWIII model data. Therefore, the abnormality index was not used for CFW forecasting based on WWIII model data.

3.2. Model Training

A total of 63 CFW accidents were recorded in the study area. For the training set, 40 of these waves were selected, and the remaining 23 formed the validation set. The training set was used for training the BPNN model to predict both CFW and non-CFW conditions. Although only 40 sea states leading to CFW accidents were available, numerous sea states without CFW occurrence (i.e., calm sea states) were existing. Therefore, 40 conditions without CFW occurrence were selected to represent these sea states for model training. To ensure that no CFWs occurred in these selected conditions, the following criteria were established for identifying calm sea states: significant wave height, peak wave period, and wind speed within the first quartile. That is, significant wave height less than 0.6 m, peak wave period less than 6.9 s, and wind speed less than 2.6 m/s were considered to be indicative of a calm sea state without CFWs. On the basis of these criteria, 40 conditions without CFW occurrence were randomly selected from the database. Finally, a data set of size 80 (40 instances with CFW and 40 instances without CFW) was established for training the BPNN model.

For BPNN model training, a supervised learning algorithm was used to assign inputs and outputs to the BPNN model during the training phase. The inputs were the variables detailed in Section 3.1 obtained from both buoy observations and WWIII model data, and the outputs were probability values. Typically, the outputs would be set as 1 and 0, indicating the occurrence and absence of a CFW, respectively. However, preliminary analysis revealed that this approach causes BPNN predictions to tend toward 1 and 0. In this circumstance, the BPNN model resembles a classification model that categorizes the occurrence or absence of a CFW rather than a regression model that predicts the probability of CFW occurrence. Therefore, the output values of the training set were modified to increase the flexibility and practicality of the probabilistic forecasting model. For the 40 CFWs, output values were randomly assigned in the interval of 0.70–0.99. Similarly, for the 40 conditions without CFWs, output values were randomly assigned between 0.01 and 0.30. Thus, a probability higher than 0.7 was regarded as a high probability of CFW occurrence, and a probability smaller than 0.3 was regarded as a low probability of CFW occurrence. The input data were normalized to the interval from 0 to 1 on the basis of the minima and maxima for the variables in the training data set.

Instability introduced by the random selection of the 40 conditions without CFWs and the random assignment of output probability values was accounted for by performing the training process 10 times. That is, a 10-fold cross validation method was applied during model training. Moreover, because of the temporal arrangement of input and output variables, two BPNN models with forecasting lead times of 12 and 24 h were developed. The BPNN model based on buoy data was termed Model I, and that using WWIII model data was termed Model II. Both models had network architectures with five hidden nodes and the hyperbolic tangent sigmoid transfer function serving as the activation function.

3.3. Model Performance

The proposed BPNN model forecasts the probability of CFW occurrence in the subsequent 12 or 24 h period. Although the model forecasts are probability values, the practical result is information regarding the presence or absence of a CFW. Therefore, the model performance could be determined using the evaluation scheme typically used for classification models. When the BPNN model performs predictions, a probability greater than or equal to 0.7 is considered to predict a CFW occurrence and a probability less than or equal to 0.3 is considered to predict that no CFW will occur. Table 1 presents an example confusion matrix for the CFW forecasting results. The rows of the matrix represent the predicted class, and the columns represent the actual class. In Table 1, the model predicts A + B CFW events, where A represents the correctly forecasted cases and B represents the incorrectly forecasted cases. Similarly, C represents cases in which the model predicts no CFW but a CFW did occur, and D represents cases in which the model predicts no CFW occurrence and none occurred. Thus, the model performance could be assessed on the basis of the indices of accuracy and recall.

$$\text{Accuracy} = (A + D)/(A + B + C + D) \tag{4}$$

$$\text{Recall} = A/(A+C) \tag{5}$$

Accuracy is the ratio of correct predictions to the total number of predictions and is used to assess the overall correctness of prediction results. Recall is the ratio of correctly predicted CFW events to the total number of CFW events and is more suitable than accuracy for assessing the prediction of rare events such as CFWs.

Table 2 lists forecasting results of the BPNN models for the training data. Table 3 lists the forecasting performance of the models in terms of accuracy and recall. The results of Models I and II were similar. The 24 h forecasting results were slightly inferior to the 12 h forecasting results. Overall, the probabilistic CFW forecasting models were well trained, and model performance with the training data was satisfactory.

Table 1. Example confusion matrix for coastal freak wave (CFW) forecasting results.

		Actual	
		CFW	**No CFW**
Predicted	CFW	A	B
	No CFW	C	D

Note: A, B, C, and D are the numbers of CFW or non-CFW cases.

Table 2. Training results of the CFW forecasting models.

Lead Time		Model I		Model II	
		CFW	**No CFW**	**CFW**	**No CFW**
12 h	CFW	39	1	37	1
	No CFW	1	39	3	39
24 h	CFW	36	4	37	1
	No CFW	4	36	3	39

Table 3. Forecasting performance in terms of accuracy and recall (training data).

Lead Time	Model I		Model II	
	Accuracy	**Recall**	**Accuracy**	**Recall**
12 h	98%	98%	95%	95%
24 h	90%	90%	93%	93%

4. Model Validation and Operational Forecasting Results

4.1. Model Validation

The trained probabilistic CFW forecasting models were validated using the validation data set. A total of 23 CFW events and 23 randomly selected non-CFW events (46 events in total) were used for model validation. Figure 3 presents the forecasting results of these 46 events for the two models and two lead times. Model I outperformed Model II because most of the CFW and non-CFW events were correctly predicted by Model I. The majority of CFW events were correctly predicted, but the prediction of non-CFW events was less accurate. Table 4 lists the accuracy and recall results for the validation data. For 12 h forecasting, Model I outperformed Model II. However, for 24 h forecasting, Models I and II demonstrated similar performance. The accuracy and recall rates were mostly higher than 72% (except for the accuracy of Model II for 12 h forecasting), indicating that the trained models were able to forecast CFW occurrence. Although the model validation is satisfactory, the proposed model could be improved to attain higher accuracy and recall rates as long as more CFW data are collected. According to the results for the training and validation data sets, Model I outperformed Model II for 12 h forecasting, and Models I and II demonstrated similar performance for 24 h forecasting.

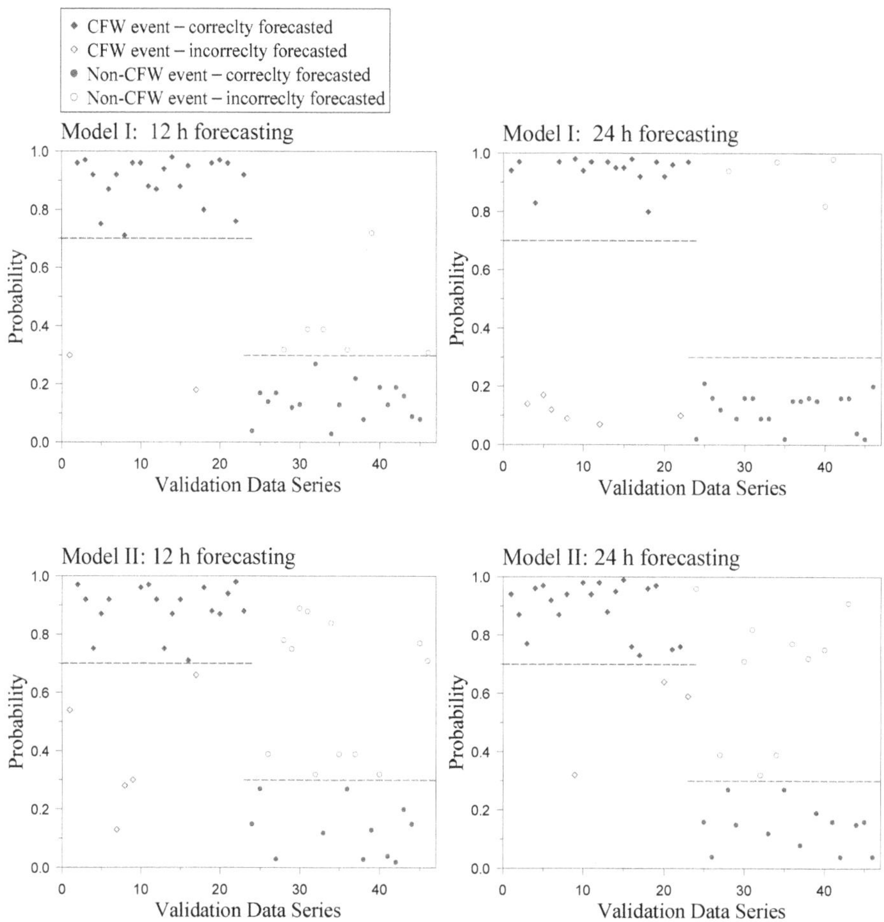

Figure 3. Probabilistic forecasting of CFW occurrence for validation data.

Table 4. Forecasting performance in terms of accuracy and recall (validation data).

Lead Time	Model I		Model II	
	Accuracy	Recall	Accuracy	Recall
12 h	83%	91%	63%	78%
24 h	78%	74%	72%	78%

4.2. Operational Probabilistic Forecasting

The proposed probabilistic CFW forecasting models were applied by the Operational Forecast System of the TCWB in experimental mode in 2018. Model I was the main model used for forecasting the probability of CFW occurrence, and Model II was a backup model used when real-time buoy observations were unavailable. The probabilistic CFW forecasting system provided probability forecasts of CFW occurrence at 6-h intervals, namely 00:00, 06:00, 12:00, and 18:00 CST. Figure 4 illustrates the forecasted probability values (rounded to the nearest integer) for August 2018. The forecasted probabilities for the lead times of 12 and 24 h were similar, but the probability yielded by 24 h forecasting was generally higher than that yielded by 12 h forecasting. The probabilistic CFW forecasting system predicted low probability of CFW occurrence during dates close to the 5th of August and at the end of the month; high probability was predicted during dates close to the 12th and 23rd of August. Some data in Figure 4 are missing because the system failed to produce forecasts because of the unavailability of real-time input data and technical problems in terms of connecting to the database. Although calculations can be performed in offline mode, the real situation of the CFW forecasting system based on its operational test in experimental mode is presented in the current study.

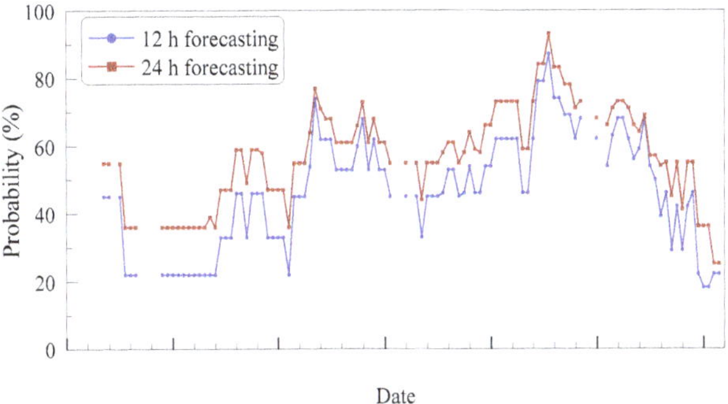

Figure 4. Probability of CFW occurrence during operational testing.

As described in Section 3, swell has a strong relationship with CFW occurrence [1,46], and the combination of swell and local wind waves may generate CFWs [47]. Therefore, the probabilistic CFW forecasting results were compared with relevant swell information. The TCWB regularly issues swell alerts when an observed or forecasted swell features a significant wave height exceeding 1.5 m and a period longer than 8 s. Thus, the swell alert data were used to assess the probabilistic CFW forecasting results. However, the TCWB does not issue swell alerts when typhoon or strong wind warnings have been issued. Therefore, swell alert data can only be compared with CFW forecasting results when typhoon or strong wind warnings have not been issued. Table 5 lists the swell alert times during 2018 and the 12 h forecasts of CFW probability for the 6 h alert period. The results indicate that the CFW forecasting system predicted high CFW probability when a swell watch was issued. As swells are related to the probability of CFW occurrence, the comparison result supports the effectiveness of the CFW forecasting system.

Table 5. Swell alert issuing time and forecasted CFW probability.

Date (yyyy/mm/dd)	Issuing Time (hh:mm)	Forecasted CFW Probability (%)
2018/05/22	16:00	80
2018/07/10	14:00	91
2018/07/22	06:00	88
2018/09/14	07:00	81
2018/09/14	16:00	82
2018/09/15	02:00	90
2018/09/15	06:00	95
2018/09/15	14:00	96
2018/09/26	01:00	95
2018/09/28	03:00	96
2018/09/28	15:00	95
2018/09/28	22:00	97
2018/09/30	23:00	94
2018/10/04	23:00	93
2018/10/06	06:00	90
2018/10/28	22:00	88

4.3. CFW Accidents during the Period of Typhoon Jebi in 2018

Typhoons introduce large swells to coastal areas and may increase the probability of CFW occurrence. During the afternoon of 2 September 2018, the weather in Taiwan was sunny and calm, but swells with a wave weight of 1.5 m were observed on the east coast. At that time, Typhoon Jebi, a category 4 typhoon according to the Saffir–Simpson hurricane wind scale, was located 1300 km east of Taiwan (Figure 5). However, four CFW accidents were reported on the east coast, with six fatalities. Three accidents occurred between 15:00 and 17:00 approximately 70–80 km south of Longdong, and one occurred at 20:00 approximately 25 km northwest of Longdong. Figure 6 shows the probabilistic forecasting of CFW occurrence in the Longdong headland during the period of Typhoon Jebi. The CFW forecasting model predicted high probability of CFW occurrence on September 2nd and 5th. At the times when the mentioned CFW accidents occurred, the forecasted probability was 86%. Although the probabilistic CFW forecasting model was developed for CFW occurrence in the Longdong headland, it can serve as a reference for CFW occurrence throughout the east coast of Taiwan. The casualties caused by CFWs during Typhoon Jebi demonstrate the necessity of the CFW forecasting model.

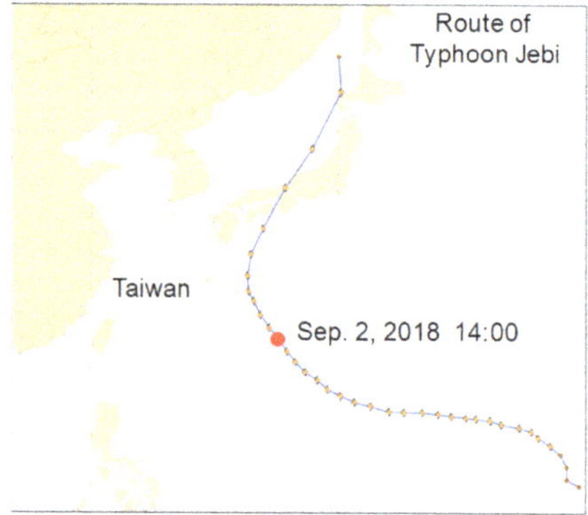

Figure 5. Route of Typhoon Jebi in 2018.

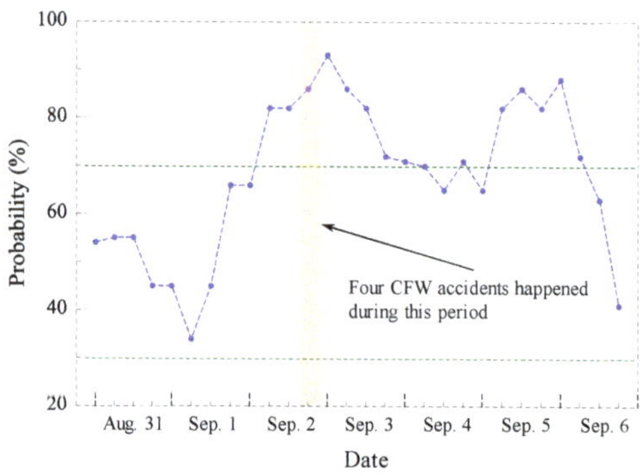

Figure 6. Probabilistic CFW forecasting during the period of Typhoon Jebi.

5. Conclusions

Although CFWs are uncommon, they have been reported worldwide. In Taiwan, CFWs have caused many accidents and casualties, mostly on the northeast coast. Forecasting the probability of CFW occurrence and disseminating warning information can mitigate the potential harm caused by CFWs. However, the CFW is difficult to simulate using physical wave models. Therefore, the present study used an AI method to forecast the occurrence of CFWs. A BPNN, which is a classical artificial neural network, was adopted to provide a probabilistic CFW forecasting model in this study.

Records of 63 CFW accidents that occurred during 2000–2016 in the Longdong headland and its neighboring coast were collected; this is the area in which CFW accidents occur most frequently in Taiwan. CFW accident data were checked and confirmed by the TCWB and were used to train and validate the forecasting model. Non-CFW data were randomly selected from calm sea states for model training and validation. A scheme was designed to assign the probability of CFW occurrence for model training. Model inputs were wave and meteorological variables with physical relationships to CFWs, namely significant wave height, peak wave period, onshore wind speed, misalignment between wind and wave directions, kurtosis of sea surface elevation, Benjamin–Feir index, groupiness factor, and abnormality index.

The input data were obtained from two sources, namely buoy observations and a wave model simulation. Therefore, two forecasting models, Models I and II, were trained using these data sources to perform probabilistic CFW forecasting. The models yield predictions at 6 h intervals with lead times of 12 and 24 h. The results of Models I and II were similar during training, but Model I outperformed Model II for the validation set in terms of accuracy and recall. Therefore, for the operational forecasting system, which was tested during 2018, Model I was used as the main model, and Model II was a backup model. The probabilistic CFW forecasting results were closely correlated with swell, which is physically related to CFWs, suggesting that the proposed model is suitable for forecasting CFW occurrence. Moreover, four CFW accidents occurred on the east coast of Taiwan on 2 September 2018, when Typhoon Jebi was located in the Pacific Ocean far away from Taiwan. The CFW forecasting model predicted a probability of 86% for CFW occurrence during the time that these accidents happened. The CFW accidents during the Typhoon Jebi period indicate the necessity of the CFW forecasting model. The probabilistic CFW forecasting model is undergoing tests and improvements and will be employed in the Operational Forecast System of the TCWB for CFW warnings.

Author Contributions: Conceptualization, D.-J.D., S.-T.C. and C.-H.T.; Data curation, Y.-C.C.; Formal analysis, D.-J.D., S.-T.C. and Y.-C.C.; Methodology, D.-J.D., S.-T.C. and C.-H.T.; Writing—original draft, D.-J.D.; Writing—review & editing, D.-J.D. and S.-T.C. All authors have read and agreed to the published version of the manuscript.

Funding: This research was supported by the Central Weather Bureau and the Ministry of Science and Technology (grant no. MOST 106-2628-E-006-008-MY3) of Taiwan.

Acknowledgments: This research was supported by the Central Weather Bureau and the Ministry of Science and Technology (grant no. MOST 106-2628-E-006-008-MY3) of Taiwan. Buoy data were measured and processed by the Coastal Ocean Monitoring Center of National Cheng Kung University. The authors express their gratitude for all support.

Conflicts of Interest: The authors declare no conflict of interest.

References

1. Chien, H.; Kao, C.C.; Chuang, L.Z. On the characteristics of observed coastal freak waves. *Coast. Eng. J.* **2002**, *44*, 301–319. [CrossRef]
2. Tsai, C.H.; Su, M.Y.; Huang, S.J. Observations and conditions for occurrence of dangerous coastal waves. *Ocean Eng.* **2004**, *31*, 745–760. [CrossRef]
3. Didenkulova, I.I.; Slunyaev, A.V.; Pelinovsky, E.N.; Kharif, C. Freak waves in 2005. *Nat. Hazards Earth Syst.* **2006**, *6*, 1007–1015. [CrossRef]
4. Didenkulova, I.; Anderson, C. Freak waves of different types in the coastal zone of the Baltic Sea. *Nat. Hazards Earth Syst.* **2010**, *10*, 2021–2029. [CrossRef]
5. Nikolkina, I.; Didenkulova, I. Rogue waves in 2006–2010. *Nat. Hazards Earth Syst.* **2011**, *11*, 2913–2924. [CrossRef]
6. Doong, D.J.; Peng, J.P.; Chen, Y.C. Development of a warning model for coastal freak wave occurrences using an artificial neural network. *Ocean Eng.* **2018**, *169*, 270–280. [CrossRef]

7. Feraco, F.; Marino, R.; Pumir, A.; Primavera, L.; Mininni, P.D.; Pouquet, A.; Rosenberg, D. Vertical drafts and mixing in stratified turbulence: Sharp transition with Froude number. *Europhys. Lett.* **2018**, *123*, 44002. [CrossRef]
8. French, M.N.; Krajewski, W.F.; Cuykendall, R.R. Rainfall forecasting in space and time using a neural network. *J. Hydrol.* **1992**, *137*, 1–31. [CrossRef]
9. Chen, S.T.; Yu, P.S.; Liu, B.W. Comparison of neural network architectures and inputs for radar rainfall adjustment for typhoon events. *J. Hydrol.* **2011**, *405*, 150–160. [CrossRef]
10. Lin, G.F.; Jhong, B.C.; Chang, C.C. Development of an effective data-driven model for hourly typhoon rainfall forecasting. *J. Hydrol.* **2013**, *495*, 52–63. [CrossRef]
11. Jhong, Y.D.; Chen, C.S.; Lin, H.P.; Chen, S.T. Physical hybrid neural network model to forecast typhoon floods. *Water* **2018**, *10*, 632. [CrossRef]
12. Dariane, A.B.; Azimi, S. Streamflow forecasting by combining neural networks and fuzzy models using advanced methods of input variable selection. *J. Hydroinform.* **2018**, *20*, 520–532. [CrossRef]
13. Bray, M.; Han, D. Identification of support vector machines for runoff modeling. *J. Hydroinform.* **2004**, *6*, 265–280. [CrossRef]
14. Yu, P.S.; Chen, S.T.; Chang, I.F. Support vector regression for real-time flood stage forecasting. *J. Hydrol.* **2006**, *328*, 704–716. [CrossRef]
15. Han, D.; Chan, L.; Zhu, N. Flood forecasting using support vector machines. *J. Hydrol.* **2007**, *9*, 267–276. [CrossRef]
16. Chen, S.T.; Yu, P.S. Real-time probabilistic forecasting of flood stages. *J. Hydrol.* **2007**, *340*, 63–77. [CrossRef]
17. Chen, S.T.; Yu, P.S. Pruning of support vector networks on flood forecasting. *J. Hydrol.* **2007**, *347*, 67–78. [CrossRef]
18. Chen, S.T.; Yu, P.S.; Tang, Y.H. Statistical downscaling of daily precipitation using support vector machines and multivariate analysis. *J. Hydrol.* **2010**, *385*, 13–22. [CrossRef]
19. Chen, S.T. Multiclass support vector classification to estimate typhoon rainfall distribution. *Disaster Adv.* **2013**, *6*, 110–121.
20. Chen, S.T. Mining informative hydrologic data by using support vector machines and elucidating mined data according to information entropy. *Entropy* **2015**, *17*, 1023–1041. [CrossRef]
21. Yu, P.S.; Chen, C.J.; Chen, S.J. Application of gray and fuzzy methods for rainfall forecasting. *J. Hydrol. Eng.* **2000**, *5*, 339–345. [CrossRef]
22. Yu, P.S.; Chen, S.T.; Wu, C.C.; Lin, S.C. Comparison of grey and phase-space rainfall forecasting models using fuzzy decision method. *Hydrol. Sci. J.* **2004**, *49*, 655–672. [CrossRef]
23. Yu, P.S.; Chen, S.T.; Chen, C.J.; Yang, T.C. The potential of fuzzy multi-objective model for rainfall forecasting from typhoons. *Nat. Hazards* **2005**, *34*, 131–150. [CrossRef]
24. Yu, P.S.; Chen, S.T. Updating real-time flood forecasting using a fuzzy rule-based model. *Hydrol. Sci. J.* **2005**, *50*, 265–278. [CrossRef]
25. Chen, C.S.; Jhong, Y.D.; Wu, T.Y.; Chen, S.T. Typhoon event-based evolutionary fuzzy inference model for flood stage forecasting. *J. Hydrol.* **2013**, *490*, 134–143. [CrossRef]
26. Lohani, A.K.; Goel, N.K.; Bhatia, K.K.S. Improving real time flood forecasting using fuzzy inference system. *J. Hydrol.* **2014**, *509*, 25–41. [CrossRef]
27. Chen, C.S.; Jhong, Y.D.; Wu, W.Z.; Chen, S.T. Fuzzy time series for real-time flood forecasting. *Stoch. Environ. Res. Risk Assess.* **2019**, *33*, 645–656. [CrossRef]
28. Deo, M.C.; Sridhar Naidu, C. Real time wave forecasting using neural networks. *Ocean Eng.* **1998**, *26*, 191–203. [CrossRef]
29. Deo, M.C.; Jha, A.; Chaphekar, A.S.; Ravikant, K. Neural networks for wave forecasting. *Ocean Eng.* **2001**, *28*, 889–898. [CrossRef]
30. Tsai, C.P.; Lin, C.; Shen, J.N. Neural network for wave forecasting among multi-stations. *Ocean Eng.* **2002**, *29*, 1683–1695. [CrossRef]
31. Mandal, S.; Prabaharan, N. Ocean wave forecasting using recurrent neural networks. *Ocean Eng.* **2006**, *33*, 1401–1410. [CrossRef]
32. Mahjoobi, J.; Mosabbeb, E.A. Prediction of significant wave height using regressive support vector machines. *Ocean Eng.* **2009**, *36*, 339–347. [CrossRef]

33. Elbisy, M.S. Sea wave parameters prediction by support vector machine using a genetic algorithm. *J. Coast. Res.* **2013**, *31*, 892–899. [CrossRef]

34. Kazeminezhad, M.H.; Etemad-Shahidi, A.; Mousavi, S.J. Application of fuzzy inference system in the prediction of wave parameters. *Ocean Eng.* **2005**, *32*, 1709–1725. [CrossRef]

35. Özger, M.; Şen, Z. Prediction of wave parameters by using fuzzy logic approach. *Ocean Eng.* **2007**, *34*, 460–469. [CrossRef]

36. Mahjoobi, J.; Etemad-Shahidi, A.; Kazeminezhad, M.H. Hindcasting of wave parameters using different soft computing methods. *Appl. Ocean Res.* **2008**, *30*, 28–36. [CrossRef]

37. Chang, H.K.; Chien, W.A. A fuzzy–neural hybrid system of simulating typhoon waves. *Coast. Eng.* **2006**, *53*, 737–748. [CrossRef]

38. Chang, H.K.; Chien, W.A. Neural network with multi-trend simulating transfer function for forecasting typhoon wave. *Adv. Eng. Softw.* **2006**, *37*, 184–194. [CrossRef]

39. Tsai, C.C.; Wei, C.C.; Hou, T.H.; Hsu, T.W. Artificial neural network for forecasting wave heights along a ship's route during hurricanes. *J. Waterw. Port Coast.* **2017**, *144*, 04017042. [CrossRef]

40. Wei, C.C. Nearshore wave predictions using data mining techniques during typhoons: a case study near Taiwan's northeastern coast. *Energies* **2017**, *11*, 11. [CrossRef]

41. Chen, S.T. Probabilistic forecasting of coastal wave height during typhoon warning period using machine learning methods. *J. Hydroinform.* **2019**, *21*, 343–358. [CrossRef]

42. Doong, D.J.; Chen, C.C.; Kao, C.C.; Lee, B.C.; Yeh, S.P. Data quality check procedures of an operational coastal ocean monitoring network. *Ocean Eng.* **2007**, *34*, 234–246. [CrossRef]

43. Rumelhart, D.E.; Hinton, G.E.; Williams, R.J. Learning representations by back-propagating errors. *Nature* **1986**, *323*, 533–536. [CrossRef]

44. Haykin, S. *Neural Networks: A Comprehensive Foundation*; MacMillan: New York, NY, USA, 1994.

45. Hagan, M.T.; Demuth, H.B.; Beale, M.H. *Neural Network Design*; PWS Publishing: Boston, MA, USA, 1996.

46. Tamura, H.; Waseda, T.; Miyazawa, Y. Freakish sea state and swell-windsea coupling: Numerical study of the Suwa-Maru incident. *Geophys. Res. Lett.* **2009**, *36*, L01607. [CrossRef]

47. Gramstad, O.; Trulsen, K. Can swell increase the number of freak waves in a wind sea? *J. Fluid Mech.* **2010**, *650*, 57–79. [CrossRef]

48. Mori, N.; Janssen, P.A. On kurtosis and occurrence probability of freak waves. *J. Phys. Oceanogr.* **2006**, *36*, 1471–1483. [CrossRef]

49. Benjamin, T.B.; Feir, J.E. The disintegration of wave trains on deep water Part 1. Theory. *J. Fluid Mech.* **1967**, *27*, 417–430. [CrossRef]

50. Janssen, P.A. Nonlinear four-wave interactions and freak waves. *J. Phys. Oceanogr.* **2003**, *33*, 863–884. [CrossRef]

51. Mori, N.; Onorato, M.; Janssen, P.A. On the estimation of the kurtosis in directional sea states for freak wave forecasting. *J. Phys. Oceanogr.* **2011**, *41*, 1484–1497. [CrossRef]

52. Funke, E.R.; Mansard, E.P.D. On the synthesis of realistic sea states. *Coast. Eng. Proc.* **1980**, *1*, 2974–2991.

53. Goda, Y. *Random Seas and Design of Maritime Structures*; World Scientific Publishing Company: Singapore, 2000.

54. Dean, R. Freak waves: A possible explanation. In *Water Wave Kinematics*; Torum, A., Gudmestat, O., Eds.; Kluwer: Dordrecht, The Netherlands, 1990; pp. 609–612.

55. Kharif, C.; Pelinovsky, E.; Slunyaev, A. *Rogue Waves in the Ocean*; Springer Science & Business Media: Berlin/Heidelberg, Germany, 2008.

Journal of
*Marine Science
and Engineering*

Article

Numerical Simulation of Large Wave Heights from Super Typhoon Nepartak (2016) in the Eastern Waters of Taiwan

Shih-Chun Hsiao [1], Hongey Chen [2,3], Han-Lun Wu [1], Wei-Bo Chen [2,*], Chih-Hsin Chang [2], Wen-Dar Guo [2], Yung-Ming Chen [2] and Lee-Yaw Lin [2]

[1]   Department of Hydraulic and Ocean Engineering, National Cheng Kung University, Tainan City 70101, Taiwan; schsiao@mail.ncku.edu.tw (S.-C.H.); hlwu627@mail.ncku.edu.tw (H.-L.W.)

[2]   National Science and Technology Center for Disaster Reduction, New Taipei City 23143, Taiwan; hchen@ntu.edu.tw (H.C.); chang.c.h@ncdr.nat.gov.tw (C.-H.C.); wdguo@ncdr.nat.gov.tw (W.-D.G.); ymcheng@ncdr.nat.gov.tw (Y.-M.C.); yaw@ncdr.nat.gov.tw (L.-Y.L.)

[3]   Department of Geosciences, National Taiwan University, Taipei City 10617, Taiwan

*   Correspondence: wbchen@ncdr.nat.gov.tw; Tel.: +886-2-8195-8612

Received: 2 March 2020; Accepted: 17 March 2020; Published: 20 March 2020

Abstract: Super Typhoon Nepartak (2016) was used for this case study because it is the most intense typhoon that made landfall in Taiwan in the past decade. Winds extracted from the Climate Forecast System version 2 (CFSV2) and ERA5 datasets and merged with a parametric typhoon model using two hybrid techniques served as the meteorological conditions for driving a coupled wave-circulation model. The computed significant wave heights were compared with the observations recorded at three wave buoys in the eastern waters of Taiwan. Model performance in terms of significant wave height was also investigated by employing the CFSV2 winds under varying spatial and temporal resolutions. The results of the numerical experiments reveal that the simulated storm wave heights tended to decrease significantly due to the lower spatial resolution of the hourly winds from the CFSV2 dataset; however, the variations in the storm wave height simulations were less sensitive to the temporal resolution of the wind field. Introducing the combination of the CFSV2 and the parametric typhoon winds greatly improved the storm wave simulations, and similar phenomena can be found in the exploitation of the ERA5 dataset blended into the parametric wind field. The overall performance of the hybrid winds derived from ERA5 was better than that from the CFSV2, especially in the outer region of Super Typhoon Nepartak (2016).

Keywords: storm wave height; super typhoon; wave-circulation model; hybrid winds

---

1. Introduction

Typhoons are usually associated with extreme wind waves and storm surges, impacting navigational safety, infrastructure in nearshore and offshore waters (harbors, seawalls, lighthouses, etc.), and coastal habitats [1–5]. The oceanic and coastal hazards caused by typhoon-driven storm waves pose a greater threat to human life, property, and infrastructure than storm surges in Taiwan. This is particularly true when a super typhoon makes landfall in Taiwan. A typhoon is designated as a "super typhoon" when its wind speeds near the typhoon center exceed 114 kt (equivalent to 58 m/s) and its intensity reaches category 4 or 5 on the Saffir–Simpson scale [6]. A super typhoon ranks among the most destructive natural hazards worldwide because it produces high storm surges, large storm waves, and torrential downpours [7]. For instance, a seawall at a fishing port near the southeastern coastal waters of Taiwan was broken in 2018 by large storm waves from several typhoons (as shown in the upper-right panel of Figure 1); a lighthouse located in a fishing port in the southeastern offshore waters was destroyed by Super Typhoon Meranti-induced extreme storm waves in 2016 (as shown in the lower-right panel of Figure 1). Therefore, accurately predicting, simulating and hindcasting typhoon-induced storm wave heights is important for the prevention and mitigation of disasters in the ocean and along coasts [2–4]. Assessments of typhoon-generated extreme wave heights with different return periods are very useful for designing seawalls to protect boats, ships, nuclear power plants, and coastal critical infrastructures from wave impacts. Moreover, the strength evaluation of oil platforms and wind turbine supporting systems is highly dependent on simulating extreme waves accurately [8,9].

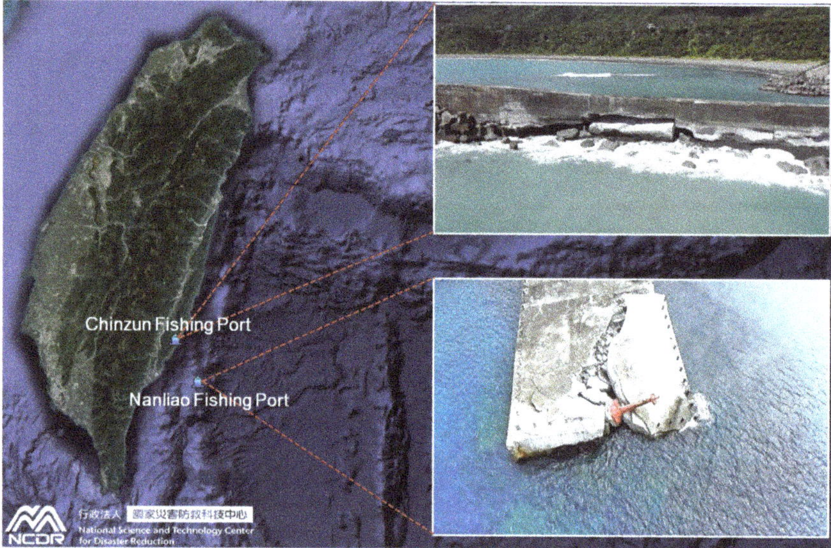

Figure 1. Typhoon-driven storm waves breached a sea wall (upper-right panel) and damaged a lighthouse (lower-left panel) in the southeastern waters of Taiwan (photo by the present study).

A comprehensive understanding of the ocean surface waves arising from extreme weather conditions is of great interest to coastal and ocean engineers and oceanographers. Many studies about predicting, simulating, or hindcasting typhoon-generated storm waves and storm surges have been carried out through numerical models because wave buoys cannot be deployed throughout an entire marine area [10–14]. A third-generation spectral wind wave model is mainly driven by wind fields and has been widely used to predict, simulate, and hindcast wave heights. Therefore, accurate wind forcing data are essential for predicting sea states with great accuracy, especially with extreme waves created by typhoons [4]. Moreover, improvements in wind forcing for wave-circulation models are believed to enhance the performances of large wave and large surge simulations [4,5,15].

The objective of the present study is to assess the effects of wind sources from different spatial and temporal resolutions on storm wave height simulation and to investigate the optimal hybrid typhoon winds through two approaches for the best performance of storm wave height simulation during Super Typhoon Nepartak in 2016. The paper is organized as follows: Section 2 describes the typhoon, observational data, bathymetry, wind forcing, and coupled wave-circulation model that were used in the present study. Section 3 presents the comparisons between the simulations and measurements for the significant wave height (SWH) time series using different wind fields. Section 4 provides a discussion of the results from a series of numerical experiments. Finally, a summary and conclusions are presented in Section 5.

2. Data and Methodology

2.1. Super Typhoon Nepartak (2016)

Super Typhoon Nepartak was the third most intense tropical cyclone worldwide in 2016. It initialized in the southern waters of Guam on 30 June and was subsequently named Nepartak by the Japan Meteorological Agency (JMA) after it had intensified into a tropical cyclone on 3 July. The Joint Typhoon Warning Center (JTWC) of the United States upgraded Nepartak to a category 4-equivalent super typhoon on 5 July and estimated that Nepartak had become a category 5-equivalent super typhoon the next day. Super Typhoon Nepartak reached its peak intensity approximately 835 km east-southeast of Taitung County, Taiwan, on 6 July; meanwhile, the JMA evaluated 10-min maximum sustained winds of 205 km/h and a central pressure of 900 hPa. Super Typhoon Nepartak made landfall in Taitung County and subsequently moved into the Taiwan Strait from Tainan County on 8 July. After that, Typhoon Nepartak continued moving inland into China as a tropical depression and dissipated on 10 July. According to the report from the Central Weather Burau (CWB) of Taiwan, Meranti was the most intense typhoon that had made landfall in Taiwan in the past 10 years (2010–2019). Figure 2 demonstrates the track and central pressure (intensity) of Super Typhoon Nepartak (2016).

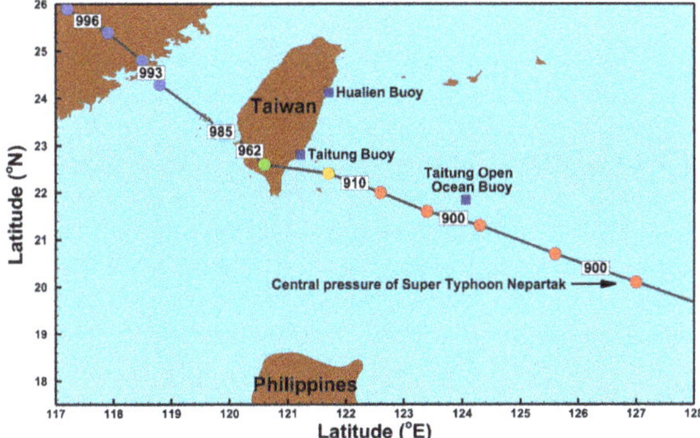

Figure 2. Track and central air pressures of Super Typhoon Nepartak in 2016 and the locations of wave buoys. The color of the solid circles represents the typhoon intensity as classified on the Saffir–Simpson scale (red: category 5; yellow: category 4; green: category 3; cyan: category 2; blue: category 1).

2.2. Wave Buoy

The hourly SWH measurements recorded at three wave buoys by the CWB, Taiwan, in the eastern waters of Taiwan during Super Typhoon Nepartak (2016) were acquired to evaluate the optimal wind field for simulating typhoon-driven SWHs. The wave buoys record the SWH, wave direction and mean wave period, and wind speed and direction at 2 m height above sea level with a sampling frequency of 2 Hz, and an accuracy of ±10 cm for the SWH. The locations of the three wave buoys, namely, the Hualien, Taitung, and Taitung Ocean buoys, are shown in Figure 2, and their corresponding coordinates and water depths are listed in Table 1.

Table 1. Information on wave buoys.

Buoy Name	Longitude (°E)	Latitude (°N)	Water Depth (m)
Taitung Ocean	124.0742	21.7664	5610
Taitung	121.1450	22.7240	30
Hualien	121.6314	24.0319	21

Data source: The CWB of Taiwan.

2.3. Description and Configuration of the Wave-Circulation Model

In the present study, the hydrodynamics of the waters surrounding Taiwan were simulated from 1 to 20 July 2016, by means of the semi-implicit cross-scale hydroscience integrated system model (SCHISM). The SCHISM was developed by Zhang et al. [16] with many improvements and enhancements from the original semi-implicit Eulerian–Lagrangian finite-element (SELFE) model, [17]. The SCHISM employs finite-element, finite-volume methods and hydrostatic and Boussinesq approximations to solve the shallow water equations. The SCHISM eliminates the Courant–Friedrichs–Lewy (CFL) stability restrictions due to the application of the semi-implicit schemes for all governing equations. The numerical stability constraints are further relaxed because the nonlinear advection terms in the momentum equations are treated with a Eulerian–Lagrangian method. The introduction of the no-mode-splitting technique in the SCHISM contributes to eliminating numerical errors regarding the splitting between internal and external modes [18]. SCHISM and SELFE have been widely used for simulating the inundation associated with storm surge [19–21], assessing the effects of sea-level rise on tidal current energy [22] and predicting flash floods in mountainous areas [23]. Typhoon-induced hydrodynamics can be well mimicked by depth integration, i.e., a 2D model; moreover, a 2D model requires fewer computing resources and less model execution time than a 3D model. Thus, a 2D model is preferred to a 3D model for storm surge, storm tide, and wind wave modeling, and the depth-integrated version of the SCHISM, SCHISM-2D, was used in the present study. A time step of 120 s and a Manning coefficient of 0.025 were set in the SCHISM-2D following [4,15], which are based on the numerical stability of the SCHISM-2D and the type of sea bottom material in Taiwanese waters.

The third-generation spectral wave model, called the wind wave model version III (WWM-III), was adopted in the present study since phase-averaged spectral wave models are capable of predicting and simulating sea state variations in the ocean [24]. Roland [25] overhauled not only the physics and numerical schemes of the WWM-II but also the computing efficiency of the model. The WWM-III solves the wave action equation on an unstructured grid by adopting the fractional step method proposed by [26]. The minimum and maximum directions for the simulation are $0°$ and $360°$, respectively, while the number of directional bins is 36. The lowest and highest frequency limits of the discrete wave period are 0.03 and 1.0 Hz, respectively, which are partitioned into 36 frequency bins. The constant for computing waves breaking in shallow water areas is 0.78. The peak enhancement of 3.3 is according to the report from the Joint North Sea Wave Project (JONSWAP, [27]), and the bottom friction coefficient is set as 0.067 in the WWM-III. An efficient computation can be achieved through the employment of different time steps for hydrodynamic and wind wave models in a wave-circulation coupled modeling system, and therefore, a time step of 600 s was assigned to the WWM-III. This means that SCHISM-2D delivers the wind and depth-integrated velocities and water surface elevations to WWM-III; in addition, the wave radiation stresses for SCHISM-2D are used by the WWM-III at five hydrodynamic time intervals. The SCHISM-WWM-III has been successfully applied to generate potential storm wave hazards for Taiwan [2], evaluate the effect of the wind field on wind wave hindcasting [4], and quantify the nonlinear interactions of typhoon-induced storm tides. Details on the coupling procedures for the SCHISM-WWM-III can be found in [28].

The computational domain must be large enough to simulate storm waves and storm surges generated by typhoons traveling long distances from east to west of Taiwan. In the present study, coverage from $105°E$ to $140°E$ and $15°N$ to $31°N$ was created for the computational domain of the SCHISM-WWM-III. This area is composed of 276,639 unstructured grids and 540,510 triangular elements; coarse meshes with 20–40 km resolutions were arranged in the open ocean beyond the coastal region, while fine meshes with 200–400 m resolutions were distributed along the coastline of Taiwan and its offshore islands (as shown in Figure 3a). Using a higher resolution mesh is fundamental to an accurate simulation; however, the computing demand will increase as the mesh becomes finer. Additionally, the surf zones and shallow waters are well characterized with the mesh resolution developed in the present study (Figure 3b) according to the report from [29–31].

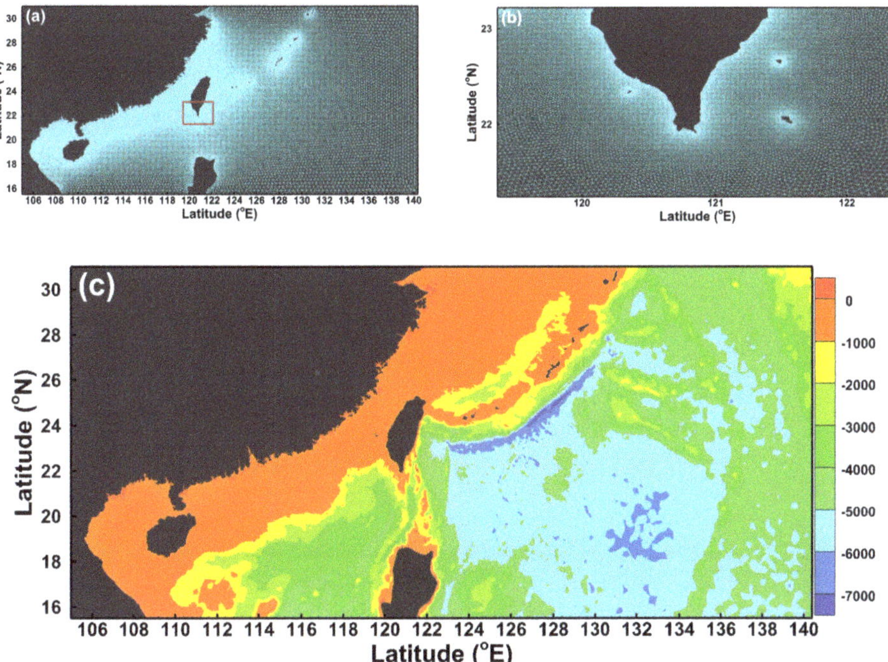

Figure 3. Entire computational meshes (**a**), enlargement of high-resolution unstructured grids for the southern coastal waters of Taiwan (**b**), and bathymetry (**c**).

A global-scale and a local-scale dataset were incorporated into gridded bathymetric data for the SCHISM-WWM-III. The GEBCO-2019 grid is the latest global-scale bathymetric product, with a spatial resolution of 15 arcsecond, released by the General Bathymetric Chart of the Oceans (GEBCO) and has been developed through the Nippon Foundation-GEBCO Seabed 2030 Project. The local-scale dataset has a coverage from 100°E to 128°E and from 4°N to 29°N with a 200-m spatial resolution provided by the Department of Land Administration and the Ministry of the Interior in Taiwan. The combined two datasets were interpolated to represent the bottom elevation in the SCHISM-WWM-III (as shown in Figure 3c). The SCHISM-2D and WWM-III take advantage of sharing the same numerical mesh to enhance the computing efficiency.

2.4. Tidal and Atmospheric Forcing

2.4.1. Tidal Forcing

Eight main tidal constituents (M_2, S_2, N_2, K_2, K_1, O_1, P_1, and Q_1) were extracted from a regional inverse tidal model (China Seas and Indonesia, [32]) and served as the tidal boundary conditions in the SCHISM-WWM-III. Although the present study focused on wind wave simulation, an inverted barometer effect resulting from the variations in atmospheric pressure was also considered in the model.

2.4.2. CFSV2 Wind Field

The Climate Forecast System version 2 (CFSV2) is an upgrade of the Climate Forecast System (CFS) and was developed by the Environmental Modeling Center at the National Centers for Environmental Prediction (NCEP). The CFSV2 is a fully coupled model representing the interactions among the ocean, land, sea ice, and atmosphere with advanced physics, increased resolution, and refined initialization to improve the seasonal climate forecasts, which makes the CFSV2 superior to the CFS. Additionally, the same model was used in the CFSV2 and the NCEP CFS reanalysis (CFSR). The CFSV2 was implemented at NCEP as the real-time seasonal forecast system in March 2011 [33]; hence, NCEP started running the CFSV2 operationally to produce analyses and forecasts. The CFSV2 products are available at horizontal resolutions of 0.205°, 0.5°, 1.875°, and 2.5° with different time intervals.

2.4.3. ERA5 Reanalysis Wind Field

ERA5 is the fifth-generation, latest global climate atmospheric reanalysis from the European Centre for Medium-Range Weather Forecasts (ECMWF). ERA5 is based on a recent ECMWF model cycle that includes coupling with ocean waves and a land model and utilizes the 4D-Var atmospheric assimilation method. ERA5 is created at a considerably higher resolution than ERA-Interim, thus replacing the widely used ERA-Interim reanalysis soon. The global hourly ERA5 atmospheric parameters are available from the Climate Data Store over latitude–longitude grids with a 0.25° by 0.25° resolution and over 137 levels in the vertical direction. The hourly ERA5 provides a more detailed evolution of extreme severe weather events and daily updates within 3 months of real time. ERA5 now covers the period from 1979 to the present; however, it will be extended from 1950 to the present by the mid-2020s.

2.4.4. Parametric Typhoon Wind Field

Many analytical parametric typhoon models have been developed to construct or reconstruct the wind and air pressure fields of typhoons for simulating storm waves and surges due to their simplicity [34–40]. The modified Rankine vortex (MRV) typhoon wind model proposed by [37] provides a more accurate wind field [41,42], which is satisfactory for ocean wave and storm surge modeling [4,37,43,44] and was therefore employed in the present study. A shape parameter α is used in the MRV model to adjust the distribution of typhoon wind speed (W) in the radial direction,

$$W = \begin{cases} W_{\max}\left(\frac{r}{R_{\max}}\right)^{\alpha} & \text{for } r < R_{\max} \\ W_{\max}\left(\frac{R_{\max}}{r}\right)^{\alpha} & \text{for } r \geq R_{\max} \end{cases} \tag{1}$$

where $W_{\max}$ is the maximum wind speed, $R_{\max}$ is the radius of maximum wind speed, and α is specified as 0.5 based on the results from [4,40,42]. Atkinson and Holliday [45] suggested an empirical relationship for $W_{\max}$ and the central pressure of typhoons (P_c):

$$W_{\max} = 3.44(1010 - P_c)^{0.644} \tag{2}$$

The radius of maximum wind speed ($R_{\max}$) can be described as a function of $W_{\max}$ and the latitude of the typhoon's center (φ):

$$R_{\max} = 46.4\exp(-0.155 \times W_{\max} + 0.0169 \times \varphi) \tag{3}$$

2.4.5. Hybrid Typhoon Wind Field

Combining the reanalysis or dynamic and parametric typhoon winds is believed to generate a better wind field in both the near-field and far-field regions of typhoons for storm wave and surge modeling [1,4,5,15,46–48]. Hence, two methods were applied to construct the hybrid winds from the CFSV2 and MRV and from the ERA5 and MRV. Method number 1 (H1) blends CFSV2 (or ERA5) and MRV directly through the following formula [1]:

$$W_{H1} = \begin{cases} W_P & r < 2R_{\max} \\ \beta^{0.70\beta^{0.06}}W_P + (1-\beta)^{0.72(1-\beta)^{0.28}}W_R & 2R_{\max} \leq r \leq 7R_{\max} \\ W_R & r > 7R_{\max} \end{cases} \tag{4}$$

where W_{H1} is the hybrid wind speed computed via Equation (4) and W_P and W_R are parametric typhoon and reanalysis winds, respectively. $\beta = (7 - r/R_{\max})/5$, in which r is the radial distance from the center of the typhoon to an arbitrary point.

Method number 2 (H2) is recommended by [46] and can be expressed as follows:

$$R_1 = R_{OP} - \gamma R_T \tag{5}$$

$$R_2 = R_{OP} + (1 - \gamma)R_T \tag{6}$$

$$W_{H2} = \begin{cases} W_P & r < R_1 \\ (1 - \lambda)W_P + \lambda W_R & R_1 \leq r \leq R_2 \\ W_R & r > R_2 \end{cases} \tag{7}$$

where R_{OP} is the optimal radius with a minimum mean difference between W_P and W_R; R_T is the width of the transition zone and γ is the empirical coefficient. For comparison with H1, R_T and γ were given as 5 $R_{\max}$ and 0.0, respectively, in the present study. $\lambda = (r - R_1)/(R_2 - R_1)$. Details for the estimation of R_{OP} are available in [46].

2.5. Model Performance Metrics

The model capacity for the simulation of the SWH time series during Super Typhoon Nepartak (2016) is quantified via three statistical indicators, i.e., the scatter index (*SI*), correction coefficient (*CC*) and *HH* (Hanna and Heinold [49]) indicators. Mentaschi et al. [50] suggested that the *HH* indicator provides more reliable and accurate statistical information for assessing the accuracy of numerical models. Thus, in addition to *SI* and *CC*, the *HH* indicator was adopted for evaluation of model performance. The three criteria are defined as:

$$SI = \sqrt{\frac{\frac{1}{n}\sum_{i=1}^{n}\left[\left(S_i - \overline{S}\right) - \left(O_i - \overline{O}\right)\right]^2}{\overline{O}}} \tag{8}$$

$$CC = \frac{\sum_{i=1}^{n}\left(S_i - \overline{S}\right)\left(O_i - \overline{O}\right)}{\sqrt{\sum_{i=1}^{n}\left(S_i - \overline{S}\right)^2}\sqrt{\sum_{i=1}^{n}\left(O_i - \overline{O}\right)^2}} \tag{9}$$

$$HH = \sqrt{\frac{\sum_{i=1}^{m}(S_i - O)^2}{\sum_{i=1}^{n}S_iO_i}} \tag{10}$$

where n is the total number of data points, S_i is the simulation, O_i is the observation, and $\overline{S}$ and $\overline{O}$ are the mean values of the simulations and observations, respectively.

3. Results

The winds from the CFSV2 and ERA5, as well as winds from a combination of an MRV model with the CFSV2 and ERA5 through the different techniques were exerted on the SCHISM-WWM-III to simulate the SWHs during Super Typhoon Nepartak (2016). The effects of the spatial and temporal resolutions of the wind field on storm wave height simulation, and the model performance using the different hybrid wind fields were investigated by conducting several numerical experiments.

3.1. Effects of the Spatial Resolutions of Wind Fields on Storm Wave Simulation

The hourly CFSV2 winds with horizontal resolutions of 0.205°, 0.5°, and 1.875° were interpolated into the SCHISM-WWM-III. The instantaneous wind field distributions are illustrated in Figure 4a–c at resolutions of 0.205°, 0.5°, and 1.875°. Wind speeds from the hourly CFSV2 with a spatial resolution of 0.205° could reach 45–50 m/s around the near-field region of Super Typhoon Nepartak (2016) (Figure 4a); however, the wind speeds decreased dramatically when a coarser spatial resolution was used. The wind speeds of the hourly CFSV2 only reached 35–40 m/s (Figure 4b) and 20–30 m/s (Figure 4c) at 0.5° and 1.875° resolutions, respectively. Furthermore, the wind field obtained from the hourly CFSV2 with a lower spatial resolution shows a more random distribution (Figure 4b,c). The computed instantaneous SWHs corresponding to hourly CFSV2 winds with different spatial resolutions are demonstrated in Figure 4d–f. Storm waves exceeding 20 m driven by hourly CFSV2 winds at spatial resolutions of 0.205° and 0.5° could be found in the deep ocean (Figure 4d,e). Wave heights were only 8–10 m at the same time if imposing the hourly CFSV2 winds at horizontal resolution of 1.875° on the SCHISM-WWM-III (Figure 4c). Figure 5a–c shows the model-data comparisons for the SWH time series from the Taitung Ocean (Figure 5a), Taitung (Figure 5b), and Hualien (Figure 5c) buoys employing the winds from the hourly CFSV2 with different spatial resolutions. Similar to the results in Figure 4d–f, hourly CFSV2 winds at a higher spatial resolution led to the greater overestimation of storm wave heights. This phenomenon is obvious at the wave buoys near the track of Super Typhoon Nepartak (2016), i.e., the Taitung Ocean and Taitung buoys.

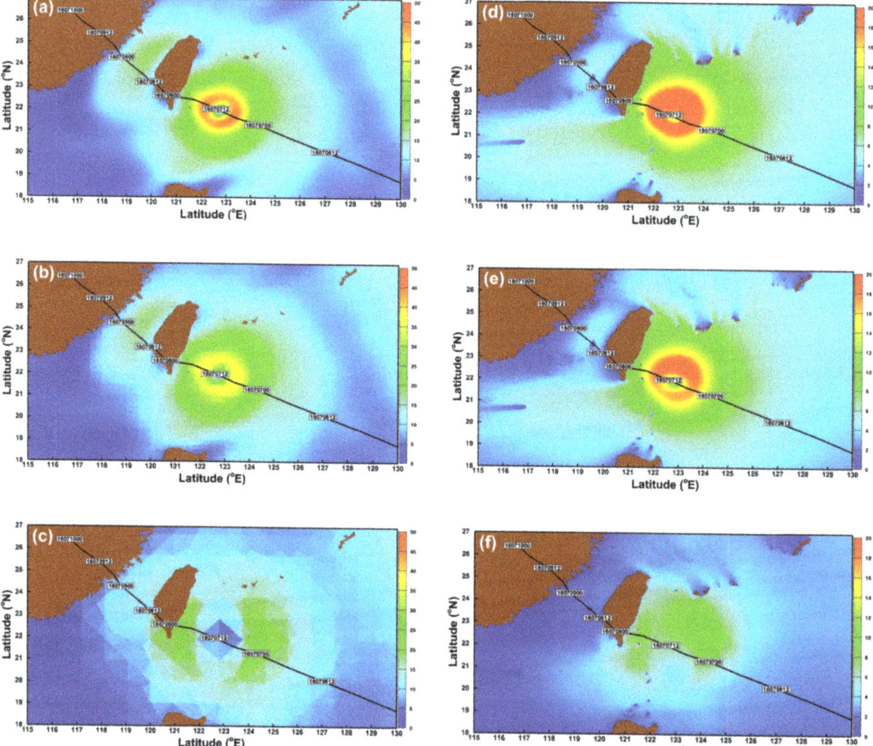

Figure 4. Instantaneous wind field distribution from the hourly Climate Forecast System version 2 (CFSV2) with spatial resolutions of (**a**) 0.205°, (**b**) 0.5°, and (**c**) 1.875° and their corresponding SWHs (**d–f**) for Super Typhoon Nepartak at 11:00 on 12 July in 2016.

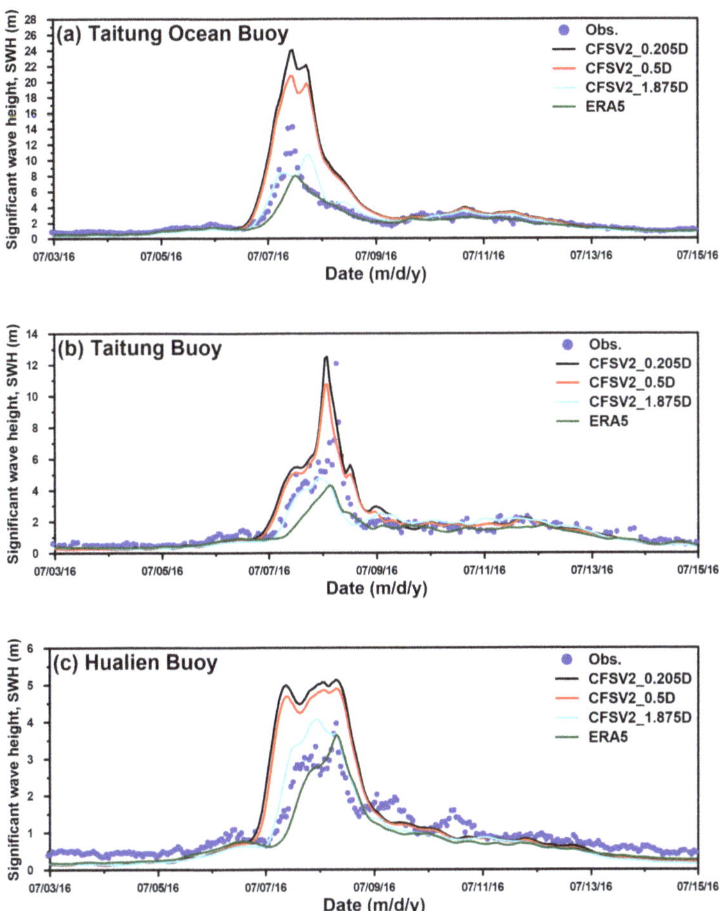

Figure 5. Model-data comparison for SWH time series at the (**a**) Taitung Ocean, (**b**) Taitung, and (**c**) Hualien buoys employing the winds from the hourly CFSV2 at different spatial resolutions.

3.2. Effects of the Temporal Resolutions of Wind Fields on Storm Wave Simulation

To better understand the effects of wind temporal resolutions on wave modeling performance, wind fields from CFSV2 with a spatial resolution of 0.205° and temporal resolutions of 1, 2, 3, and 6 h were utilized for reproducing the SWHs during Super Typhoon Nepartak (2016). The instantaneous wind field distributions from CFSV2 with the same spatial resolution (0.205°) and different temporal resolutions are depicted in Figure 6a–d for temporal resolutions of 1, 2, 3, and 6 h, respectively. The variations in wind field from higher to lower temporal resolutions are insignificant except for a delay in time. Figure 6e–h expresses the distributions of simulated instantaneous SWHs corresponding to Figure 6a–d. The patterns of storm waves are insensitive to variations in the wind temporal resolution. Figure 7a–c shows the comparisons of the SWH time series between the modeled and measured values for the Taitung Ocean (Figure 7a), Taitung (Figure 7b), and Hualien (Figure 7c) buoys using winds from the 0.205° CFSV2 with different temporal resolutions. The comparisons shown in Figure 7a–c are dissimilar to those in Figure 5a–c. The storm wave simulations from different wind temporal resolutions are almost identical, except that a 2-m difference was present at the Taitung buoy if the spatial resolution of the wind field remained unchanged.

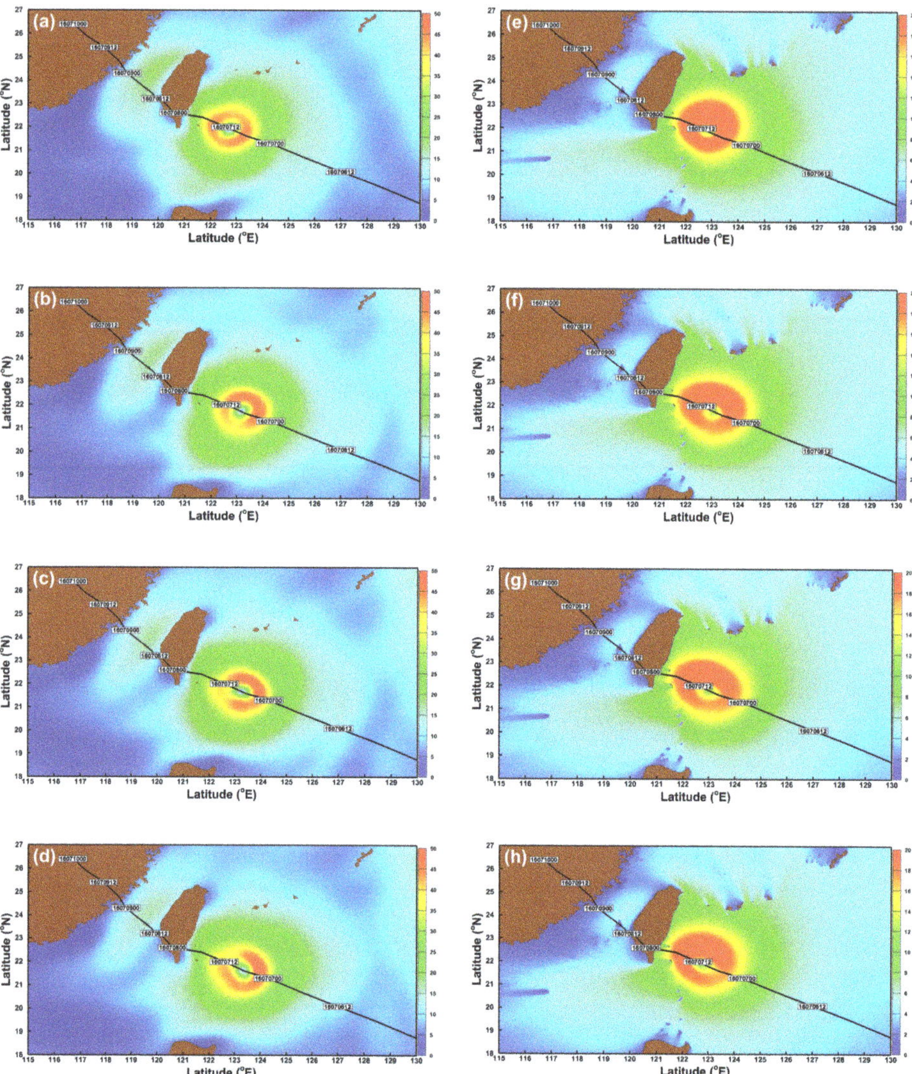

Figure 6. Instantaneous wind field distribution from CFSV2 with a spatial resolution of 0.205° and temporal resolutions of (**a**) 1 h, (**b**) 2 h, (**c**) 3 h, and (**d**) 6 h and their corresponding SWHs (**e–h**) during Super Typhoon Nepartak at 11:00 on 12 July in 2016.

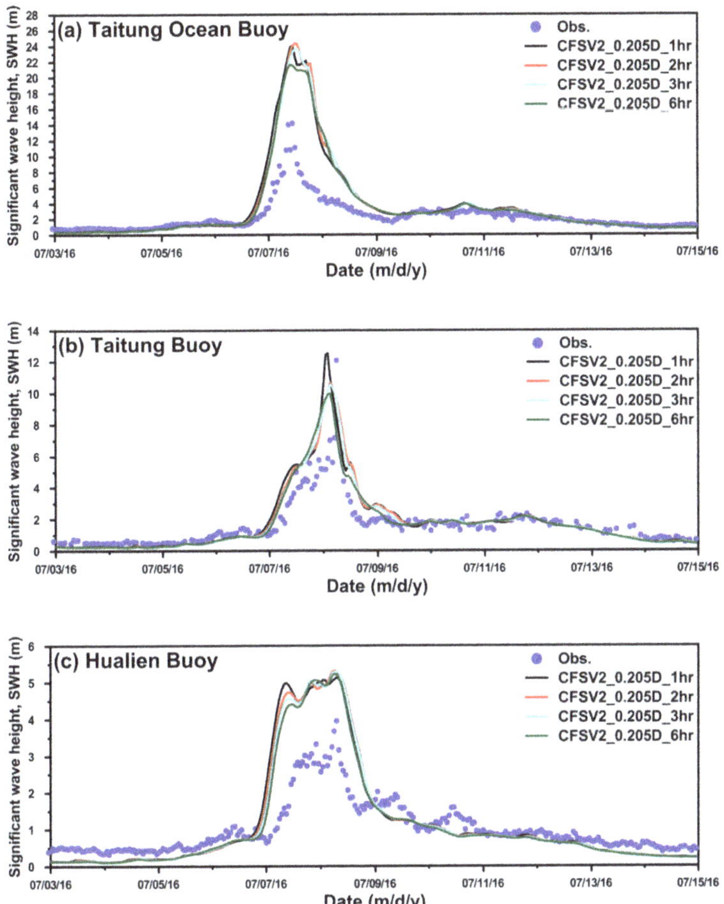

Figure 7. Model-data comparison for the SWH time series of the (**a**) Taitung Ocean, (**b**) Taitung, and (**c**) Hualien buoys employing the winds from CFSV2 with a spatial resolution of 0.205° and different temporal resolutions.

3.3. Effects of Hybrid Wind Fields on Storm Wave Simulation Through Different Superposed Techniques

Although the storm wave heights were overestimated using atmospheric forcing from the hourly CFSV2 with a spatial resolution of 0.205° (hereafter CFSV2_0.025D_1hr), CFSV2_0.025D_1hr could produce a more acceptable wind field distribution. Hence, the SWH computations were again calculated by employing the wind fields derived from merging CFSV2_0.025D_1hr and the MRV typhoon model by means of Equation (4) (hereafter CFSV2_0.025D_1hr_H1) and Equation (7) (hereafter CFSV2_0.025D_1hr_H2). The instantaneous wind field distributions from CFSV2_0.025D_1hr, CFSV2_0.205D_1hr_H1 and CFSV2_0.205D_1hr_H2 are represented in Figure 8a–c. As shown in Figure 8b,c, the region with the higher wind speed (45–50 m/s) decreased and was more concentrated around the center of Super Typhoon Nepartak (2016). Figure 8d–f illustrates the spatial distributions of the SWH reflecting wind forcing from CFSV2_0.025D_1hr, CFSV2_0.205D_1hr_H1 and CFSV2_0.205D_1hr_H2. The extent of extreme waves (16–20 m) decreased, which is particularly true when utilizing winds from CFSV2_0.205D_1hr_H2 (Figure 8f). Figure 9a–c clarifies the improvements in the SWH simulations using the hybrid typhoon winds for the Taitung Ocean (Figure 9a), Taitung (Figure 9b), and Hualien (Figure 9c) buoys. A mismatch between the computed and measured maximum SWH could be reduced from over 9 m to less than 2 m at the Taitung Ocean buoy for Super Typhoon Nepartak (2016).

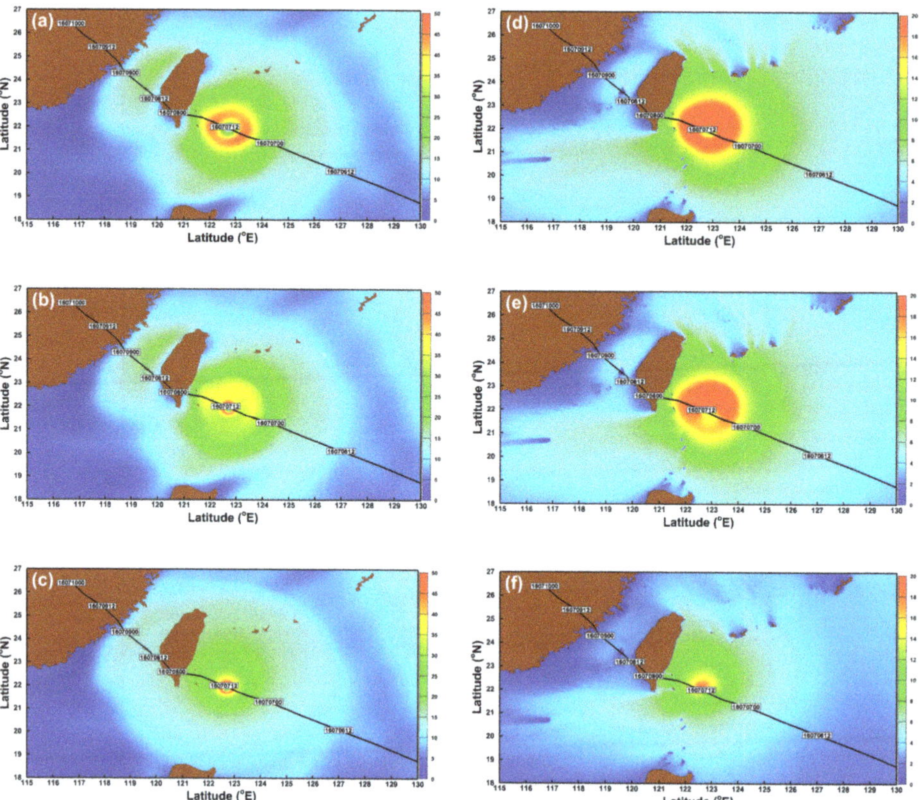

Figure 8. Instantaneous wind field distribution from the (**a**) hourly CFSV2 with a spatial resolution of 0.205°, (**b**) CFSV2_0.205D_1hr_H1, and (**c**) CFSV2_0.205D_1hr_H2 and their corresponding SWHs (**d–f**) for Super Typhoon Nepartak at 11:00 on 12 July in 2016.

The hourly reanalysis winds with a spatial resolution of 0.25° (hereafter ERA5_1hr) during Super Typhoon Nepartak (2016) obtained from the ECMWF were also blended with the MRV typhoon model via Equation (4) (hereafter ERA5_1hr_H1) and Equation (7) (hereafter ERA5_1hr_H2). Three wind fields were imposed on the SCHISM-WWM-III to reconfirm that the use of hybrid typhoon winds will greatly improve the storm wave height simulations. Similar to Figure 8a–f, the instantaneous wind fields of ERA5_1hr, ERA5_1hr_H1 and ERA5_1hr_H2 and their corresponding SWH distributions are displayed in Figure 10a–f. The wind speeds of the original ERA5_1hr for Super Typhoon Nepartak (2016) were extremely weak (<30 m/s, as shown in Figure 10a) and consequently led to a smaller storm wave height (<10 m, as shown in Figure 10d). The SWHs of Super Typhoon Nepartak (2016) driven by ERA5_1hr winds were even smaller than those induced by hourly CFSV2 winds with a spatial resolution of 1.875° (as shown green lines in Figure 5a–c). Exerting higher speed winds from ERA5_1hr_H1 and ERA5_1hr_H2 (40–45 m/s, as shown in Figure 10b,c) on SCHISM-WWM-III could cause larger storm waves (approximately 16 m, as shown in Figure 10e,f). Figure 11a–c depicts a comparison of the SWH time series between the simulations and measurements for the Taitung Ocean, Taitung, and Hualien buoys. Introducing a combination of the MRV typhoon wind and reanalysis product into a wave-circulation model significantly enhances the storm wave simulations in both the inner (Figure 11a,b) and outer (Figure 11c) regions of the typhoon. However, ERA5_1hr_H1 resulted in better SWH simulations than ERA5_1hr_H2 for the Hualien buoy (Figure 11c).

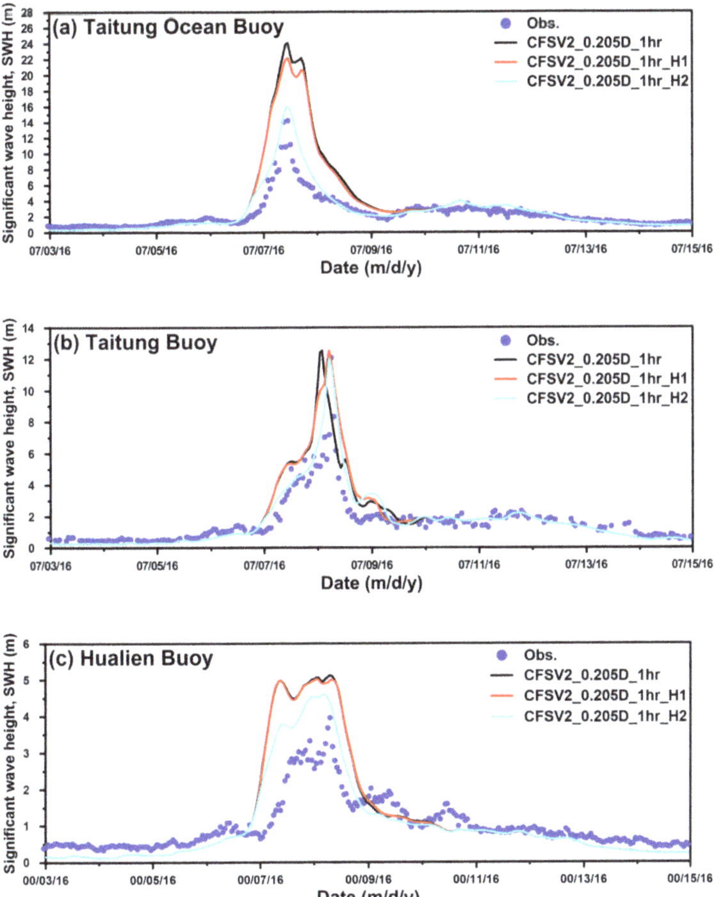

Figure 9. Model-data comparison for the SWH time series the (**a**) Taitung Ocean, (**b**) Taitung, and (**c**) Hualien buoys employing winds from CFSV2 with a spatial resolution of 0.205°, CFSV2_0.205D_1hr_H1, and CFSV2_0.205D_1hr_H2.

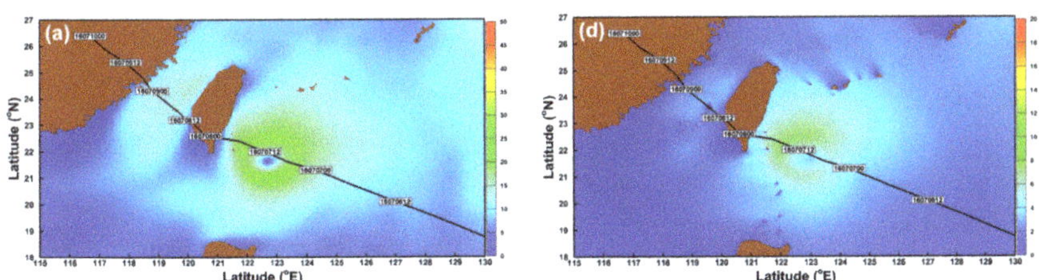

Figure 10. *Cont.*

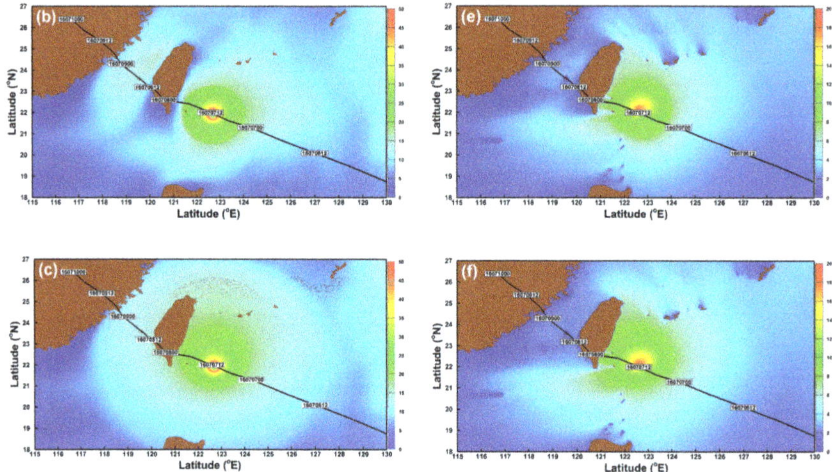

Figure 10. Instantaneous wind field distribution from the (**a**) hourly ERA5, (**b**) ERA5_1hr_H1, and (**c**) ERA5_1hr_H2 and their corresponding SWHs (**d**–**f**) for Super Typhoon Nepartak at 11:00 on 12 July in 2016.

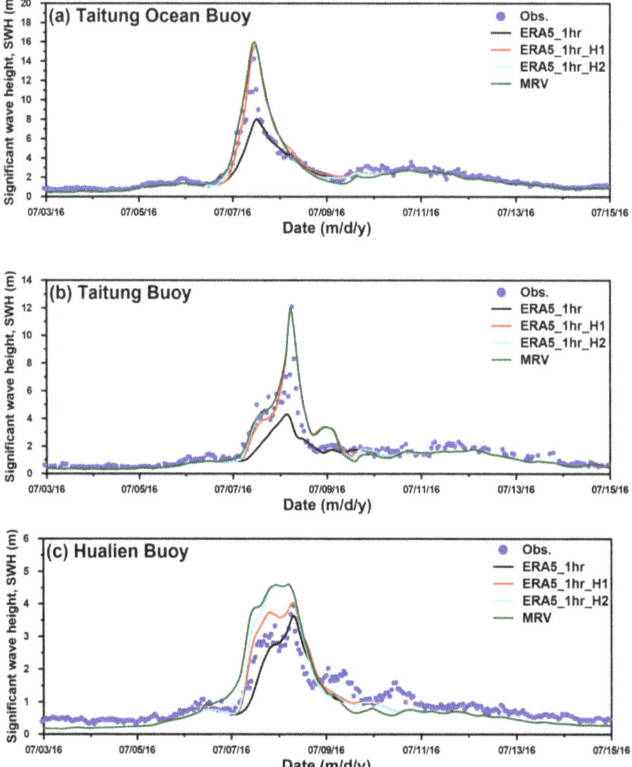

Figure 11. Model-data comparison for the SWH time series at the (**a**) Taitung Ocean, (**b**) Taitung, and (**c**) Hualien buoys employing the winds from ERA5_1hr, ERA5_1hr_H1, ERA5_1hr_H2, and modified Rankine vortex (MRV).

4. Discussion

The numerical experiments conducted in the present study show that the storm wave height simulations in both nearshore and offshore areas benefit from inputting the integrations of CFSV2 or ERA5 winds with the MRV typhoon model. The statistical analysis was implemented to quantify the simulation errors when using the CFSV2_0.205D_1hr_H2 and ERA5_1hr_H1 winds during Super Typhoon Nepartak (2016). Table 2 lists the SWH statistical parameters based on a model-data comparison for the Taitung Ocean, Taitung, and Hualien buoys, adopting winds from CFSV2_0.205D_1hr_H2. The minimum *SI* and *HH* and the maximum *CC* are 0.47, 0.32, and 0.97 for the Taitung Ocean buoy, while the maximum *SI* and *HH* and the minimum *CC* are 0.61, 0.45, and 0.92 for the Hualien buoy, respectively. The same statistical parameters are also listed in Table 3 for the utilization of winds from ERA5_1hr_H1. The overall performance of SWH simulations by the ERA5_1hr_H1 winds is slightly better than that by the CFSV2_0.205D_1hr_H2 winds. Figure 12a–f describes the scatter plots of computed SWHs versus measured SWHs for the Taitung Ocean (Figure 12a,d), Taitung (Figure 12b,e), and Hualien (Figure 12c,f) buoys with wind forcing from CFSV2_0.205D_1hr_H2 (Figure 12a–c) and ERA5_1hr_H1 (Figure 12d–f). The larger difference (simulation minus measurement range of 4–6 m) between the measurement of 7–11 m is detected at the Taitung Ocean and Taitung Buoys, as the differences are always less than 4 m at the Hualien Buoy. Tables 4 and 5 summarize the percentage of model-data difference for different classes and buoys due to the introduction of winds from CFSV2_0.205D_1hr_H2 and ERA5_1hr_H1, respectively. The maximum percentage with a model-data difference larger than 2 m is only 3.03% (2.27% + 0.76%, as shown Table 5) for the Taitung buoy using the ERA5_1hr_H1 winds. Thus, the ERA5_1hr_H1 wind served as an optimal meteorological forcing for simulating SWHs by the SCHISM-WWM-II for Super Typhoon Nepartak (2016).

Table 2. Statistical parameters of SWH based on model-data comparison for different buoys using winds from CFSV2_0.205D_1hr_H2.

Buoy Name	*SI*	*CC*	*HH*
Taitung Ocean	0.47	0.97	0.32
Taitung	0.50	0.94	0.33
Hualien	0.61	0.92	0.45

Table 3. Statistical parameters of SWH based on model-data comparison for different buoys using winds from ERA5_1hr_H1.

Buoy Name	*SI*	*CC*	*HH*
Taitung Ocean	0.39	0.96	0.28
Taitung	0.39	0.92	0.35
Hualien	0.36	0.95	0.31

Table 4. Percentage of model-data difference for different classes and buoys using winds from CFSV2_0.205D_1hr_H2.

Buoy Name	[−2,0)	[0,2)	[2,4)	[4,6)
Taitung Ocean	67.48	26.29	4.34	1.90
Taitung	64.02	32.95	2.27	0.76
Hualien	85.12	13.77	1.10	0.00

Table 5. Percentage of model-data difference for different classes and buoys using winds from ERA5_1hr_H1.

Buoy Name	[−2,0)	[0,2)	[2,4)	[4,6)
Taitung Ocean	77.51	19.75	1.08	1.63
Taitung	76.89	20.08	2.27	0.76
Hualien	88.71	11.29	0.00	0.00

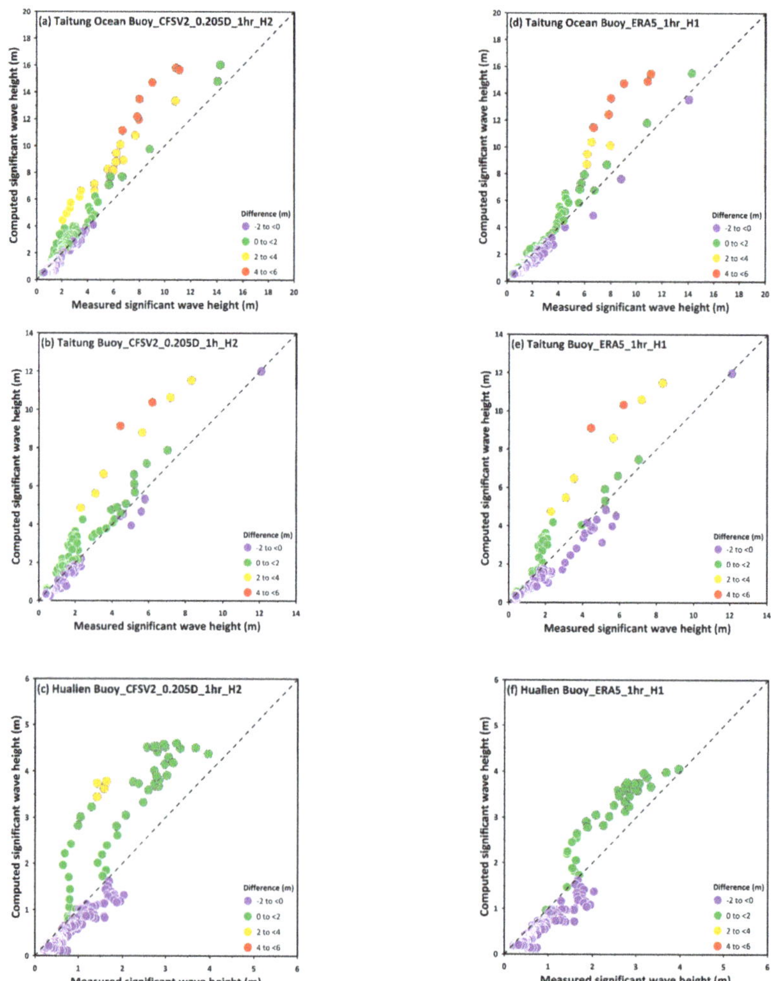

Figure 12. Scatter plot of computed versus measured SWHs at the (**a,d**) Taitung Ocean, (**b,e**) Taitung, and (**c,f**) Hualien buoys employing the winds from CFSV2_0.205D_1hr_H2 (**a–c**) and ERA5_1hr_H1 (**d–f**).

The SWH simulations driven by winds at a lower resolution are inferior to those generated by winds at a higher resolution. This phenomenon is particularly sensitive to variations in spatial resolution. The spatial distributions of the maximum SWH simulations for Super Typhoon Nepartak (2016) utilizing winds from the hourly CFSV2 with different spatial resolutions and from the CFSV2 with a spatial resolution of 0.205° and different temporal resolutions are illustrated in Figure 13a–f. The maximum SWHs attenuated obviously when the spatial resolution of the wind field was reduced from 0.205° to

1.875° (as shown in Figure 13a–c). The variations in the maximum SWH distribution are relatively small, even if the temporal resolution of the wind fields varies from hourly to 6 h (Figure 13a,d–f); moreover, the discrete maximum SWH distributions are more visible at a lower temporal resolution in the deep ocean (Figure 13e,f). Rusu et al. [51] and Van Vledder and Akpınar [52] suggested that the employment of winds with a higher spatial (mesh) resolution could improve not only typhoon wind filed but also the parameters resulting from the wave model in addition to the SWH. The statistical errors, such as bias, scatter index, and root-mean-square error, for the "SWH" and "spectral period" were usually decrease when the spatial resolution of wind is increased [52]. However, Rusu et al. [51] also recommended that the utilization of winds with a spatial resolution of 4.1 km (approximate 0.04°) is sufficient for simulating wind waves. Adopting winds in a very-high-spatial-resolution (e.g., 0.03°) atmospheric model is the best way to predict, simulate, or hindcast sea states; however, the computing demand increases. Therefore, the next best alternative is the use of a wind field resulting from the superposition of a moderate-spatial-resolution (e.g., 0.2°) atmospheric model and a parametric typhoon model.

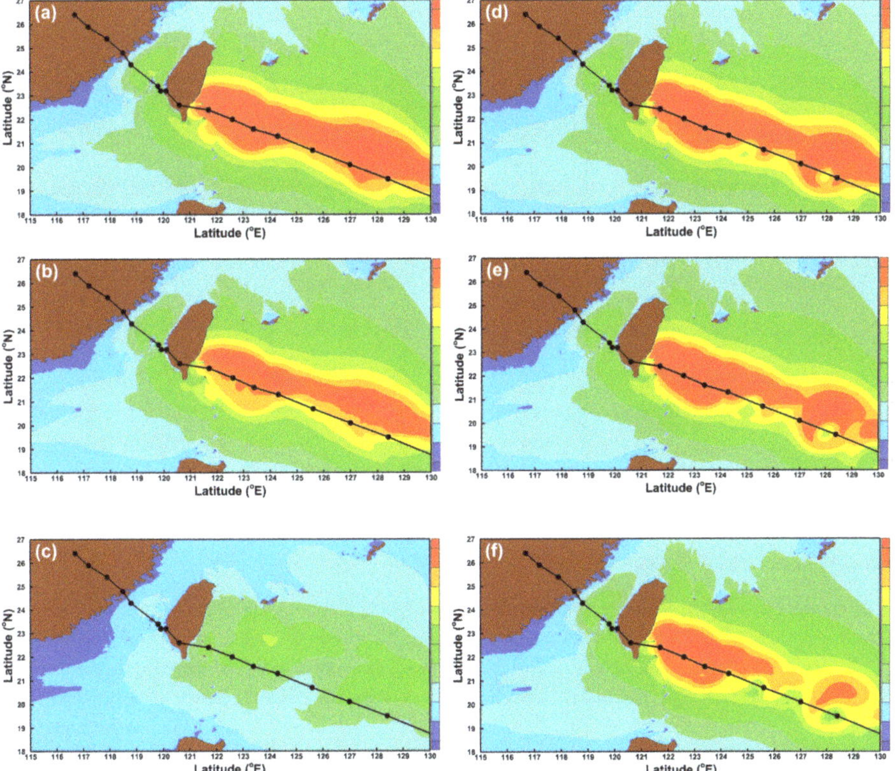

Figure 13. Spatial distribution of maximum SWHs driven by winds from the hourly CFSV2 with spatial resolutions of (**a**) 0.205°, (**b**) 0.5°, and (**c**) 1.875° and from CFSV2 with a spatial resolution of 0.205° and temporal resolutions of (**d**) 2 h, (**e**) 3 h, and (**f**) 6 h for Super Typhoon Nepartak in 2016.

Even though modeled and observed wind speed (at 10 m height above sea level) comparisons were implemented in many previous studies [5,48,51,52], the similar comparisons are difficult to conduct in the present study directly. This is because the anemometer on the Taitung Ocean, Taitung, and Hualien buoys is only 2 m above the sea level according to the report from the CWB of Taiwan.

The simulations and measurements of wave direction and mean wave period were compared in order to reconfirm the efficiency of the hybrid typhoon winds for improving the wave parameters simulation. Figure 14 shows a model-data comparison for the wave direction (Figure 14a–c) and mean wave period (Figure 14d–f) at different wave buoys. The wave direction and mean wave period simulations were improved using the hybrid typhoon winds, which is particularly true for the mean wave period simulations when winds from ERA5_1hr_H1 were exerted on the SCHISM-WWM-III. Additionally, the storm wave heights from Super Typhoon Nepartak (2016) were simulated with the MRV model alone. As shown in Figure 11a–c (green lines), the utilization of the MRV winds matched the peak SWHs well in the inner region of the typhoon (Figure 11a,b for Taitung Ocean and Taitung buoys) but overestimated the SWHs in outer regions of the typhoon (Figure 11c for Hualien buoy).

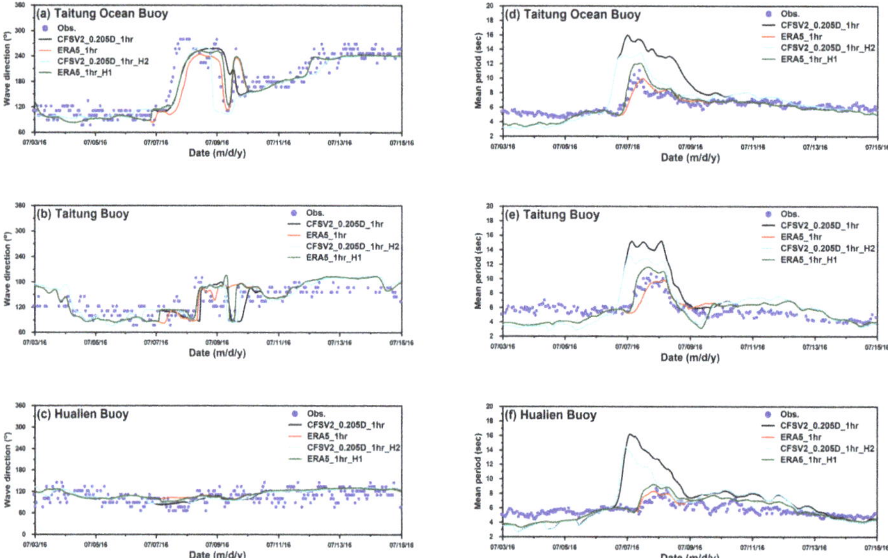

Figure 14. Model-data comparison for the wave direction (**a**–**c**) and mean wave period (**d**–**f**) time series at the (**a**,**d**) Taitung Ocean, (**b**,**e**) Taitung, and (**c**,**f**) Hualien buoys employing the winds from CFSV2_0.205D_1hr, ERA5_1hr, CFSV2_0.205D_1hr_H2, and ERA5_1hr_H1.

5. Summary and Conclusions

In this study, winds from a global atmospheric product, CFSV2, and a global atmospheric reanalysis product, ERA5, as well as winds from a combination of the parametric typhoon model with CFSV2 and ERA5 via two hybrid methods (Equations (4) and (7)), were used as meteorological conditions for a coupled wave-circulation model called SCHISM-WWM-III. The time series of simulated SWHs during Super Typhoon Nepartak (2016) were compared with the observations measured at three wave buoys in the eastern waters of Taiwan. The storm wave heights were severely overestimated due to the presence of hourly CFSV2 winds with a spatial resolution of 0.205°. The hourly CFSV2 wind fields with spatial resolutions of 0.205°, 0.5°, and 1.875° were imposed on the SCHISM-WWM-III. The winds with a coarser spatial resolution usually generated the smaller SWHs. The sensitivity of the temporal resolution of the wind field to the storm wave height simulation was also examined by adopting CFSV2 winds with a spatial resolution of 0.205° and temporal resolutions of 1, 2, 3, and 6 h. However, the simulated SWHs were insensitive to the temporal resolution of wind.

The hourly ERA5 winds with a spatial resolution of 0.25° were applied to the wave-circulation model. Wind speeds acquired from ERA5 in the inner region of Super Typhoon Nepartak (2016) were extremely weak and therefore drove the storm wave heights by only half of the measurements. To overcome the overestimation or underestimation of storm wave height simulations, hybrid typhoon winds were proposed by blending parametric typhoon models (the MRV model) and hourly CFSV2 and ERA5 with spatial resolutions of 0.205° and 0.25°, respectively, by means of two methods. The results demonstrate that using the wind fields derived from the hourly CFSV2 merged with the MRV via Equation (7) could improve the simulations of large storm wave heights close to the track of Super Typhoon Nepartak (2016) but still overestimate the wave heights in the far field of the typhoon. The computed typhoon-generated large SWHs increased significantly, especially for the inner regions of the typhoon, due to the employment of hybrid winds from the hourly ERA5 incorporated with the MRV through Equation (4). Additionally, the timings of the peak SWHs, wave direction and mean wave period during Super Typhoon Nepartak (2016) could also be corrected and well matched by utilizing hybrid wind fields.

Although this study provides insight into the spatial resolution of the CFSV2 wind being more important than the temporal resolution for SWH modeling, a reasonable distribution of typhoon-induced maximum SWHs was created by introducing a wind field with a higher temporal resolution to the SCHISM-WWM-III. Using the hybrid wind fields from CFSV2 and ERA5 produced similar storm wave heights near the track of Super Typhoon Nepartak (2016); however, the simulations of storm wave height agreed well with the observations in both the near and far fields of typhoon by applying the blended winds from ERA5 only, which is particularly essential for successful storm wave computation. The findings of this study will greatly benefit the accuracy of the operational marine weather forecast system, the design of infrastructure in nearshore waters, and the prevention of coastal disasters. Moreover, the combination of a numerical atmospheric or reanalysis product with higher spatial and temporal resolutions with parametric typhoon winds is the optimal atmospheric condition for a coupled wave-circulation model to predict or hindcast storm wave height accurately.

Author Contributions: Conceptualization, S.-C.H., H.C., H.-L.W., W.-B.C., C.-H.C., W.-D.G. and L.-Y.L. Methodology, W.-B.C. and H.-L.W. Validation, W.-B.C. Formal analysis, H.-L.W., C.-H.C., W.-D.G. and Y.-M.C. Writing—original draft preparation, W.-B.C., S.-C.H., H.-L.W. and L.-Y.L. Supervision, S.-C.H., H.C. and L.-Y.L. All authors have read and agreed to the published version of the manuscript.

Funding: This research was funded by the Ministry of Science and Technology (MOST), Taiwan, grant No. MOST 108-2625-M-865-001.

Acknowledgments: The authors would like to thank the Central Weather Bureau and Department of Land Administration, Ministry of the Interior of Taiwan for providing the measured data as well as Joseph Zhang at the Virginia Institute of Marine Science, College of William & Mary, and Aron Roland at the BGS IT&E GmbH, Darmstadt, Germany, for kindly sharing their experiences concerning the use of the numerical model.

Conflicts of Interest: The authors declare no conflict of interest.

References

1. Shao, Z.; Liang, B.; Li, H.; Wu, G.; Wu, Z. Blended wind fields for wave modeling of tropical cyclones in the South China Sea and East China Sea. *Appl. Ocean Res.* **2018**, *71*, 20–33. [CrossRef]
2. Shih, H.-J.; Chang, C.-H.; Chen, W.-B.; Lin, L.-Y. Identifying the Optimal Offshore Areas for Wave Energy Converter Deployments in Taiwanese Waters Based on 12-Year Model Hindcasts. *Energies* **2018**, *11*, 499. [CrossRef]
3. Chang, C.-H.; Shih, H.-J.; Chen, W.-B.; Su, W.-R.; Lin, L.-Y.; Yu, Y.-C.; Jang, J.-H. Hazard Assessment of Typhoon-Driven Storm Waves in the Nearshore Waters of Taiwan. *Water* **2018**, *10*, 926. [CrossRef]
4. Chen, W.-B.; Chen, H.; Hsiao, S.-C.; Chang, C.-H.; Lin, L.-Y. Wind forcing effect on hindcasting of typhoon-driven extreme waves. *Ocean Eng.* **2019**, *188*, 106260. [CrossRef]
5. Murty, P.N.L.; Srinivas, K.S.; Rao Rama Pattabhi, E.; Bhaskaran, P.K.; Shenoi, S.S.C.; Padmanabham, J. Improved cyclone wind fields over the Bay of Bengal and their application in storm surge and wave computations. *Appl. Ocean Res.* **2020**, *95*, 102048. [CrossRef]
6. Huang, F.; Xu, S. Super typhoon activity over the western North Pacific and its relationship with ENSO. *J. Ocean Univ. China* **2010**, *9*, 123–128. [CrossRef]

7. Balaguru, K.; Foltz, G.R.; Leung, L.R.; Emanuel, K.A. Global warming-induced upper-ocean freshening and the intensification of super typhoons. *Nat. Commun.* **2016**, *7*, 13670. [CrossRef] [PubMed]

8. Akpınar, A.; Bingölbali, B.; van Vledder, G.P. Wind and wave characteristics in the Black Sea based on the SWAN wave model forced with the CFSR winds. *Ocean Eng.* **2016**, *126*, 276–298. [CrossRef]

9. Kim, E.; Manuel, L.; Curcic, M.; Chen, S.S.; Phillips, C.; Veers, P. *On the Use of Coupled Wind, Wave, and Current Fields in the Simulation of Loads on Bottom Supported Offshore Wind Turbines during Hurricanes*; Technical Report NREL/TP-5000-65283; National Renewable Energy Laboratory: Golden, CO, USA, 2016.

10. Li, X.; Chen, Y.; Pan, S.; Pan, Y.; Fang, J.; Sowa, D.M.A. Estimation of mean and extreme waves in the East China Seas. *Appl. Ocean Res.* **2016**, *56*, 35–47. [CrossRef]

11. Osorio, A.F.; Montoya, R.D.; Ortiz, J.C.; Peláez, D. Construction of synthetic ocean wave series along the Colombian Caribbean Coast: A wave climate analysis. *Appl. Ocean Res.* **2016**, *56*, 119–131. [CrossRef]

12. Orimolade, A.P.; Haver, S.; Gudmestad, O.T. Estimation of extreme significant wave heights and the associated uncertainties: A case study using NORA10 hindcast data for the Barents Sea. *Mar. Struct.* **2016**, *49*, 1–17. [CrossRef]

13. Muraleedharan, G.; Lucas, C.; Soares, C.G. Regression quantile models for estimating trends in extreme significant wave heights. *Ocean Eng.* **2016**, *118*, 204–215. [CrossRef]

14. Lau, A.Y.A.; Terry, J.P.T.; Ziegler, A.D.; Switzer, A.D.; Lee, Y.; Etienne, S. Understanding the history of extreme wave events in the Tuamotu Archipelago of French Polynesia from large carbonate boulders on Makemo Atoll, with implications for future threats in the central South Pacific. *Mar. Geol.* **2016**, *380*, 174–190. [CrossRef]

15. Hsiao, S.-C.; Chen, H.; Chen, W.-B.; Chang, C.-H.; Lin, L.-Y. Quantifying the contribution of nonlinear interactions to storm tide simulations during a super typhoon event. *Ocean Eng.* **2019**, *194*, 106661. [CrossRef]

16. Zhang, Y.J.; Ye, F.; Stanev, E.V.; Grashorn, S. Seamless cross-scale modelling with SCHISM. *Ocean Model.* **2016**, *102*, 64–81. [CrossRef]

17. Zhang, Y.J.; Baptista, A.M. SELFE: A semi-implicit Eulerian-Lagrangian finite-element model for cross-scale ocean circulation. *Ocean Model.* **2008**, *21*, 71–96. [CrossRef]

18. Shchepetkin, A.F.; McWilliams, J.C. The regional oceanic modeling system (ROMS): A split-explicit, free-surface, topography-following-coordinate, oceanic model. *Ocean Model.* **2005**, *9*, 347–404. [CrossRef]

19. Wang, H.V.; Loftis, J.D.; Liu, Z.; Forrest, D.; Zhang, J. The storm surge and sub-grid inundation modeling in New York City during hurricane Sandy. *J. Mar. Sci. Eng.* **2014**, *1*, 226–246. [CrossRef]

20. Chen, W.-B.; Liu, W.-C. Modeling flood inundation Induced by river flow and storm surges over a river basin. *Water* **2014**, *6*, 3182–3199. [CrossRef]

21. Chen, W.-B.; Liu, W.-C. Assessment of storm surge inundation and potential hazard maps for the southern coast of Taiwan. *Nat. Hazards* **2016**, *82*, 591–616. [CrossRef]

22. Chen, W.-B.; Chen, H.; Lin, L.-Y.; Yu, Y.-C. Tidal Current Power Resource and Influence of Sea-Level Rise in the Coastal Waters of Kinmen Island, Taiwan. *Energies* **2017**, *10*, 652. [CrossRef]

23. Chen, Y.-M.; Liu, C.-H.; Shih, H.-J.; Chang, C.-H.; Chen, W.-B.; Yu, Y.-C.; Su, W.-R.; Lin, L.-Y. An Operational Forecasting System for Flash Floods in Mountainous Areas in Taiwan. *Water* **2019**, *11*, 2100. [CrossRef]

24. Komen, G.J.; Cavaleri, L.; Donelan, M.; Hasselmann, K.; Hasselmann, S.; Janssen, P.A.E. *Dynamics and Modelling of Ocean Waves*; Cambridge University Press: Cambridge, UK, 1994; p. 532.

25. Roland, A. Development of WWM II: Spectral Wave Modeling on Unstructured Meshes. Ph.D. Thesis, Technology University Darmstadt, Darmstadt, Germany, 2009.

26. Yanenko, N.N. *The Method of Fractional Steps*; Springer: Berlin/Hamburg, Germany, 1971.

27. Hasselmann, K.; Barnett, T.P.; Bouws, E.; Carlson, H.; Cartwright, D.E.; Enke, K.; Ewing, J.A.; Gienapp, H.; Hasselmann, D.E.; Kruseman, P.; et al. *Measurements of Wind-Wave Growth and Swell Decay during the Joint North Sea Wave Project (JONSWAP)*; Deutsches Hydrographisches Institut: Berlin/Hamburg, Germany, 1973.

28. Roland, A.; Zhang, Y.J.; Wang, H.V.; Meng, Y.; Teng, Y.-C.; Maderich, V.; Brovchenko, I.; Dutour-Sikiric, M.; Zanke, U. A fully coupled 3D wave-current interaction model on unstructured grids. *J. Geophys. Res.* **2012**, *117*, C00J33. [CrossRef]

29. Zhang, H.; Sheng, J. Estimation of extreme sea levels over the eastern continental shelf of North America. *J. Geophys. Res. Oceans* **2013**, *118*, 6253–6273. [CrossRef]

30. Muis, S.; Verlaan, M.; Winsemius, H.C.; Aerts, J.C.J.H.; Ward, P.J. A global reanalysis of storm surges and extreme sea levels. *Nat. Commun.* **2016**, *7*, 11969. [CrossRef] [PubMed]

31. Marsooli, R.; Lin, N. Numerical Modeling of Historical Storm Tides and Waves and Their Interactions Along the U.S. East and Gulf Coasts. *J. Geophys. Res. Oceans* **2018**, *123*, 3844–3874. [CrossRef]

32. Zu, T.; Gana, J.; Erofeevac, S.Y. Numerical study of the tide and tidal dynamics in the South China Sea. *Deep Sea Res.* **2008**, *55*, 137–154. [CrossRef]

33. Saha, S.; Moorthi, S.; Wu, X.; Wang, J.; Nadiga, S.; Tripp, P.; Pan, H.-L.; Behringer, D.; Hou, Y.-T.; Chuang, H.-Y.; et al. The NCEP Climate Forecast System Version 2. *J. Clim.* **2014**, *27*, 2185–2208. [CrossRef]

34. Fujita, T. Pressure distribution within typhoon. *Geophys. Mag.* **1952**, *23*, 437–451.

35. Jelesnianski, C.P. A numerical computation of storm tides induced by a tropical storm impinging on a continental shelf. *Mon. Weather Rev.* **1965**, *93*, 343–358. [CrossRef]

36. Holland, G.J. An analytical model of the wind and pressure profiles in hurricanes. *Mon. Weather Rev.* **1980**, *108*, 1212–1218. [CrossRef]

37. Phadke, A.C.; Martino, C.D.; Cheung, K.F.; Houston, S.H. Modeling of tropical cyclone winds and waves for emergency management. *Ocean Eng.* **2003**, *30*, 553–578. [CrossRef]

38. Wang, X.; Qian, C.; Wang, W.; Yan, T. An elliptical wind field model of typhoons. *J. Ocean Univ. China* **2004**, *3*, 33–39. [CrossRef]

39. MacAfee, A.W.; Pearson, G.W. Development and testing of tropical Cyclone parametric wind models tailored for midlatitude application-preliminary results. *J. Appl. Meteorol. Climatol.* **2006**, *45*, 1244–1260. [CrossRef]

40. Wood, V.T.; White, L.W.; Willoughby, H.E.; Jorgensen, D.P. A new parametric tropical cyclone tangential wind profile model. *Mon. Weather Rev.* **2013**, *141*, 1884–1909. [CrossRef]

41. Cheung, K.F.; Tang, L.; Donnelly, J.P.; Scileppi, E.M.; Liu, K.B.; Mao, K.B.; Houston, S.H.; Murnane, R.J. Numerical modeling and field evidence of coastal overwash in southern New England from Hurricane Bob and implications for paleotempestology. *J. Geophys. Res.* **2007**, *112*, F03024. [CrossRef]

42. Bastidas, L.A.; Knighton, J.; Kline, S.W. Parameter sensitivity and uncertainty analysis for a storm surge and wave model. *Nat. Hazards Earth Syst. Sci.* **2016**, *16*, 2195–2210. [CrossRef]

43. Tolman, H.L.; Alves, J.H.G.M. Numerical modeling of waves generated by tropical cyclones using moving grids. *Ocean Model.* **2005**, *9*, 305–323. [CrossRef]

44. Yu, Y.-C.; Chen, H.; Shih, H.-J.; Chang, C.-H.; Hsiao, S.-C.; Chen, W.-B.; Chen, Y.-M.; Su, W.-R.; Lin, L.-Y. Assessing the Potential Highest Storm Tide Hazard in Taiwan Based on 40-year Historical Typhoon Surge Hindcasting. *Atmosphere* **2019**, *10*, 346. [CrossRef]

45. Atkinson, G.D.; Holliday, C.R. Tropical cyclone minimum sea level pressure/maximum sustained wind relationship for the western North Pacific. *Mon. Weather. Rev.* **1977**, *105*, 421–427. [CrossRef]

46. Pan, Y.; Chen, Y.P.; Li, J.X.; Ding, X.L. Improvement of wind field hindcasts for tropical cyclones. *Water Sci. Eng.* **2016**, *9*, 58–66. [CrossRef]

47. Li, J.; Pan, S.; Chen, Y.; Fan, Y.M.; Pan, Y. Numerical estimation of extreme waves and surges over the northwest Pacific Ocean. *Ocean Eng.* **2018**, *153*, 225–241. [CrossRef]

48. Qiao, W.; Song, J.; He, H.; Li, F. Application of different wind filed models and wave boundary layer model to typhoon waves numerical simulation in WAVEWATCH III model. *Tellus A* **2019**, *7*, 1657552. [CrossRef]

49. Hanna, S.; Heinold, D. *Development and Application of a Simple Method for Evaluating Air Quality*; API Pub.: Washington, DC, USA, 1985.

50. Mentaschi, L.; Besio, G.; Cassola, F.; Mazzino, A. Problems in RMSE-based wave model validations. *Ocean Model.* **2013**, *72*, 53–58. [CrossRef]

51. Rusu, L.; Bernardino, M.; Guedes Soares, C. Influence of Wind Resolution on Prediction of Waves Generated in an Estuary. *J. Coast. Res.* **2009**, *56*, 1419–1423.

52. Van Vledder, G.P.; Akpınar, A. Wave model predictions in the Black Sea: Sensitivity to wind fields. *Appl. Ocean Res.* **2015**, *53*, 161–178. [CrossRef]

Journal of
*Marine Science
and Engineering*

Article

On the Sensitivity of Typhoon Wave Simulations to Tidal Elevation and Current

Shih-Chun Hsiao [1], Han-Lun Wu [1], Wei-Bo Chen [2,*], Chih-Hsin Chang [2] and Lee-Yaw Lin [2]

[1]   Department of Hydraulic and Ocean Engineering, National Cheng Kung University, Tainan City 70101, Taiwan; schsiao@mail.ncku.edu.tw (S.-C.H.); hlwu627@mail.ncku.edu.tw (H.-L.W.)
[2]   National Science and Technology Center for Disaster Reduction, New Taipei City 23143, Taiwan; chang.c.h@ncdr.nat.gov.tw (C.-H.C.); yaw@ncdr.nat.gov.tw (L.-Y.L.)
*   Correspondence: wbchen@ncdr.nat.gov.tw; Tel.: +886-2-8195-8612

Received: 7 September 2020; Accepted: 21 September 2020; Published: 22 September 2020

Abstract: The sensitivity of storm wave simulations to storm tides and tidal currents was investigated using a high-resolution, unstructured-grid, coupled circulation-wave model (Semi-implicit Cross-scale Hydroscience Integrated System Model Wind Wave Model version III (SCHISM-WWM-III)) driven by two typhoon events (Typhoons Soudelor and Megi) impacting the northeastern coast of Taiwan. Hourly wind fields were acquired from a fifth-generation global atmospheric reanalysis (ERA5) and were used as meteorological conditions for the circulation-wave model after direct modification (MERA5). The large typhoon-induced waves derived from SCHISM-WWM-III were significantly improved with the MERA5 winds, and the peak wave height was increased by 1.0–2.0 m. A series of numerical experiments were conducted with SCHISM-WWM-II and MERA5 to explore the responses of typhoon wave simulations to tidal elevation and current. The results demonstrate that the simulated significant wave height, mean wave period and wave direction for a wave buoy in the outer region of the typhoon are more sensitive to the tidal current but less sensitive to the tidal elevation than those for a wave buoy moored in the inner region of the typhoon. This study suggests that the inclusion of the tidal current and elevation could be more important for typhoon wave modeling in sea areas with larger tidal ranges and higher tidal currents. Additionally, the suitable modification of the typhoon winds from a global atmospheric reanalysis is necessary for the accurate simulation of storm waves over the entire region of a typhoon.

Keywords: SCHISM-WWM-III; ERA5; direct modification method; storm wave; tidal elevation; tidal current

1. Introduction

Extreme waves and storm surges resulting from typhoons are usually regarded as destructive disasters that present a great hazard to navigational safety and coastal and nearshore infrastructures and can flood low-lying areas along the coast [1–3]. To mitigate the disaster risk in both nearshore and coastal regions, many researchers have put considerable effort into the improvement of storm wave and storm surge modeling [2–4]. To achieve this goal, it is important to enhance the predictive capabilities of the typhoon storm surge and wave models used for emergency preparedness and risk evaluation. In addition, the ability to accurately predict large-wave events is critical to the success of wave energy conversion deployment [5].

Research on the nonlinear interactions between the tidal current, tidal elevation, and waves is complex because such studies require an understanding of the corresponding transfer mechanisms. The performance of typhoon storm surge and wave models strongly depends on a complete understanding of these physical processes. Longuet-Higgins and Stewart [6] and Phillips [7] clarified the transfer of momentum from ocean surface waves to tidal currents. The momentum flux between wind and current is based on the wind stress bulk formula and is proportional to the square of the wind speed [8]. The radiation stress theory proposed in Reference [6] provides an approach for introducing wave-induced forces into tidal currents. The contributions of waves to a storm surge or storm tide through radiation-stress-induced forces can be examined by coupling a circulation model with a wave model [9,10]. However, further clarification is needed to capture the contributions of the tidal current and tidal elevation to storm waves. Huang et al. [9] analyzed the effects of a storm surge on waves by comparing wave simulations with and without a storm surge for Tampa Bay in Florida. The results of their study indicated that the simulated significant wave heights could be increased by 1.0–1.5 m, and the simulated peak wave height could see perhaps as much as a 2–3 m increase. They found that the contribution of the storm surge to the significant wave height was greater than that of the significant wave height to the storm surge. This is because the Tampa Bay region is close to a gently sloping continental shelf, which allows the wind stress to be more influential than the wave-induced force in generating a storm surge. Although Huang et al. [9] investigated the storm surge effect on waves, the current understanding of the contributions of the tide and tidal current to waves is still lacking. A comprehensive understanding of the interactions between storm waves, tidal current, and tidal elevation would be of great interest to coastal and ocean engineers and oceanographers.

Rusu et al. [11] and Van Vledder and Akpınar [12] suggested that model predictions of wave fields could be greatly improved if winds of a higher spatial resolution were used in the numerical model. Rusu et al. [11] also noted that wind fields with a spatial resolution of 4.1 km are satisfactory for wind wave modeling. However, considerable computational resources and a long model execution time are required. In contrast, a simple method was applied in the present study to improve the typhoon winds acquired from a reanalysis product.

The goal of the present study was to investigate the sensitivity of storm wave simulations to tidal currents and tidal elevations during two typhoons impacting Taiwan in 2015 and 2016 by means of a series of model experiments. This paper is organized as follows. The studied typhoons, the data recorded at the wave buoys, the circulation and wave models, the bathymetry, the typhoon winds, and the boundary conditions that were used in the present study are described in Section 2. In Section 3, the significant wave height time series are compared among several different simulation scenarios. A discussion of the results from the series of model experiments and a confirmation of the results are provided in Section 4, and a summary and conclusions are offered in Section 5.

2. Data and Models

2.1. Description of the Studied Typhoons

Typhoon Soudelor, also known as Typhoon Hanna in the Philippines, was the strongest tropical cyclone of the 2015 Pacific typhoon season and the third most intense tropical cyclone worldwide in 2015. Soudelor formed as a tropical depression in the southeastern sea area of Guam on 29 July and reached its peak intensity, with a central atmospheric pressure of 900 hPa and ten-minute maximum sustained winds of 215 km/h, on 3 August. Soudelor was a Category 5-equivalent super typhoon, according to a wind speed assessment by the Joint Typhoon Warning Center (JTWC). Soudelor, which made landfall over Hualien County of Taiwan on 7 August and emerged in the Taiwan Strait early the next day, had a severe impact on Taiwan.

Typhoon Megi was a large and powerful tropical cyclone that hit Taiwan and southeastern China in late September 2016. The Japan Meteorological Agency (JMA) indicated that Typhoon Megi had reached its peak intensity at 18:00 UTC on 26 September, with ten-minute maximum sustained winds of 155 km/h and a central pressure of 940 hPa. However, Typhoon Megi had already intensified into a stronger typhoon at approximately 03:00 UTC on 27 September, with one-minute maximum sustained winds of 220 km/h (equivalent to Category 4 on the Saffir–Simpson hurricane wind scale), before it made landfall over Hualien County at 06:00 UTC. Megi emerged into the Taiwan Strait from Yunlin County at 13:10 UTC and subsequently made landfall over Fujian Province of China at 20:40 UTC on 27 September as a weaker typhoon. The tracks, central positions, and arrival times of Typhoons Soudelor and Megi were acquired from the best track data of the Regional Specialized Meteorological Center (RSMC) Tokyo-Typhoon Center, and are illustrated in Figure 1 (Figure 1a for Typhoon Soudelor and Figure 1b for Typhoon Megi).

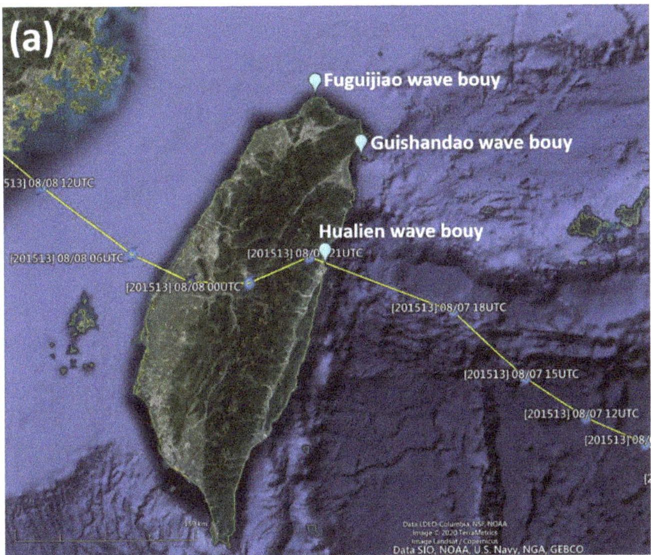

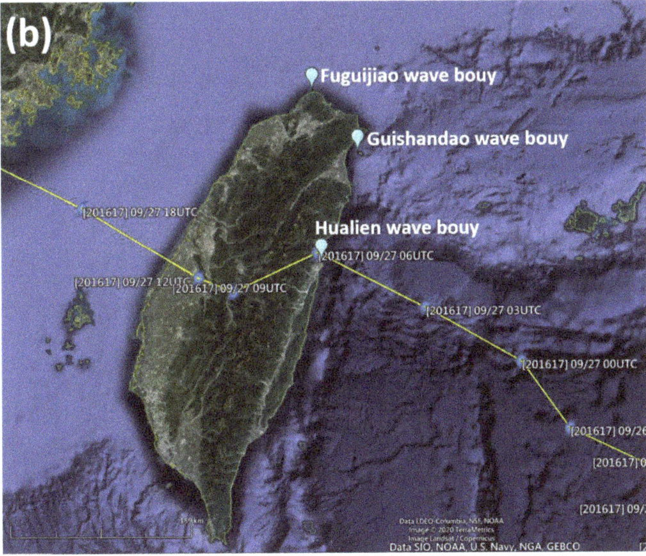

Figure 1. Tracks, central positions, and arrival times of (**a**) Typhoon Soudelor in 2015 and (**b**) Typhoon Megi in 2016. The points marked in cyan are the locations of the wave buoys adopted in the present study (Information of typhoons were provided by the Regional Specialized Meteorological Center (RSMC) Tokyo-Typhoon Center).

2.2. Description of the Adopted Wave Buoys

The Central Weather Bureau (CWB) of Taiwan has deployed many wave buoys in the nearshore waters for long-term monitoring of wave parameters. In this study, the significant wave height (Hs), mean wave period (Tm), and wave direction (WD) data measured at three wave buoys in the northeastern waters of Taiwan during Typhoon Soudelor in 2015 and Typhoon Megi in 2016 were applied to evaluate the Hs simulation performance achieved with a high-resolution coupled circulation-wave model. According to the buoy observation dada annual report from the CWB, the sampling frequency of wave buoys is 2 Hz for 10 min in the beginning of each hour. The recorded acceleration time series is transferred to acceleration spectrum and then converted to the one-dimensional wave spectrum. The zeroth (m_0) and second (m_2) moments of the wave power spectrum can be obtained by integrating the one-dimensional wave spectrum. Finally, the significant wave height and mean wave period are calculated as: Hs = $4.004\sqrt{m_0}$ and Tm = $\sqrt{(m_0/m_2)}$. The wave direction is the direction of the most energetic wave in the spectrum. The buoy names and locations are shown in Figure 1, and the coordinates and water depths of their deployment sites are listed in Table 1.

Table 1. Information on the adopted wave buoys.

Buoy Name	Longitude (°E)	Latitude (°N)	Water Depth (m)
Fuguijiao	121.535	25.3042	34
Guishandao	121.9261	24.8467	25
Hualien	121.6325	24.03111	22

Data source: The Central Weather Bureau (CWB) of Taiwan.

2.3. Barotropic 2-D Ocean Circulation Model

The Semi-implicit Cross-scale Hydroscience Integrated System Model (SCHISM) was employed to simulate the storm tidal elevations and tidal currents in the waters surrounding Taiwan during Typhoons Soudelor and Megi. In SCHISM, the primitive shallow-water equation is solved on an unstructured grid using finite-element and finite-volume methods based on the hydrostatic and Boussinesq approximations [13]. With many new corrections and upgrades from the Semi-implicit Eulerian-Lagrangian Finite-Element (SELFE) model [14], SCHISM now has an open-source distribution that is licensed under the Apache License Version 2.0. The numerical stability constraints are relaxed due to the application of the semi-implicit scheme in SCHISM. SCHISM and SELFE have been used in the simulation of storm surge inundation [15–17] and riverine flash flooding [18], in the evaluation of tidal current energy variations resulting from sea-level rise [19], and in investigation of fecal coliform transportation in an estuarine system [20]. A two-dimensional model is preferred over a three-dimensional model for storm surge, storm tide, and wind wave modeling because two-dimensional models require less execution time and computational resource consumption; therefore, SCHISM-2D, a depth-integrated version of SCHISM, was utilized in the present study to simulate the sea-surface elevations and currents. The governing equations in two-dimensional form are written in the Cartesian coordinate system as follows:

$$\frac{\partial \eta}{\partial t} + \frac{\partial uH}{\partial x} + \frac{\partial vH}{\partial y} = 0, \tag{1}$$

$$\frac{Du}{Dt} = fv - \frac{\partial}{\partial x}\left\{ g(\eta - \alpha\hat{\psi}) + \frac{P_A}{\rho_0} \right\} + \frac{\tau_{sx} + \tau_{rx} - \tau_{bx}}{\rho_0 H}, \tag{2}$$

$$\frac{Dv}{Dt} = -fu - \frac{\partial}{\partial y}\left\{ g(\eta - \alpha\hat{\psi}) + \frac{P_A}{\rho_0} \right\} + \frac{\tau_{sy} + \tau_{ry} - \tau_{by}}{\rho_0 H}, \tag{3}$$

where D is the material derivative, t is the time, $\eta(x,y,t)$ is the free-surface elevation, $u(x,y,t)$ and $v(x,y,t)$ are the horizontal velocities, h is the seafloor elevation, $H = \eta + h$ is the total water depth, f is the

Coriolis factor, g is the gravitational acceleration, $\hat{\psi}$ is the Earth's tidal potential, α is the effective Earth elasticity factor, ρ_0 is the reference water density, and $P_A(x,y,t)$ is the sea-surface atmospheric pressure. τ_{sx} and τ_{sy} are the wind stresses, which can be expressed as:

$$\tau_{sx} = \rho_a C_s \sqrt{W_x^2 + W_y^2} W_x, \tag{4}$$

$$\tau_{sy} = \rho_a C_s \sqrt{W_x^2 + W_y^2} W_y, \tag{5}$$

where ρ_a is the air density, C_s is the wind drag coefficient, and W_x and W_y are the wind speeds at 10 m above the sea surface. Powell et al. [21] suggested that an upper bound on C_s is necessary in the model in the case of high wind speeds; thus, SCHISM-2D calculates C_s following the formula proposed by Smith [22] but with an embedded cap for cases in which the wind speed exceeds a specific value:

$$C_s = 1.0^{-3} \begin{cases} 0.61 + 0.063 \times 6.0, & W < 6.0 \\ 0.61 + 0.063 \times W, & 6.0 \le W \le 50.0 \\ 0.61 + 0.063 \times 50.0, & W > 50.0 \end{cases}, \tag{6}$$

where W is the resultant wind speed.

τ_{bx} and τ_{by} in Equations (2) and (3) are the bottom shear stresses, which can be computed as:

$$\tau_{bx} = \rho_0 C_d \sqrt{u^2 + v^2} u, \tag{7}$$

$$\tau_{by} = \rho_0 C_d \sqrt{u^2 + v^2} v, \tag{8}$$

where C_d is the bottom drag coefficient, which is parameterized as follows:

$$C_d = g n^2 / H^{1/3}, \tag{9}$$

where n is the Manning coefficient. Equation (2) indicates that the bottom drag coefficient C_d varies spatially with different values of H. In Equations (2) and (3), τ_{rx} and τ_{ry} are the radiation stresses, which are computed using a wind wave model as reported by Longuet-Higgins and Stewart [6,23].

2.4. Ocean Surface Wave Model

Komen [24] suggested that the ocean surface waves induced by wind can be simulated through phase-averaged spectral wave models. Therefore, a third-generation spectral wave model, the Wind Wave Model version III (WWM-III), was utilized in the present study to model typhoon-generated extreme sea states in Taiwanese waters. Roland [25] overhauled the original WWM-I source code from Hsu et al. [26] and upgraded it to WWM-II. Comparisons between WWM-II and the Simulating WAves Nearshore (SWAN) model, developed by Booij et al. [27], for test cases of continental shelf refraction showed that the Residual Distribution Flux Corrected Transport scheme implemented in WWM-II is able to accurately resolve the directional spreading over a continental shelf profile and that the resulting simulations are far superior to those using the SWAN model with a higher-order scheme at the same mesh resolution. WWM-III is the latest version of WWM, with many enhancements and improvements, and has already been applied to reproduce the long-term wave parameters in the nearshore waters of Taiwan [28]. In WWM-III, the governing wave action balance equation is solved on an unstructured grid:

$$\frac{\partial N}{\partial t} + \frac{\partial (C_{gx} + u)N}{\partial x} + \frac{\partial (C_{gy} + v)N}{\partial y} + \frac{\partial (C_\sigma N)}{\partial \sigma} + \frac{\partial (C_\theta N)}{\partial \theta} = \frac{S_{tot}}{\sigma}, \tag{10}$$

where $N(\sigma, \theta)$ is the spectral wave action density, C_{gx} and C_{gy} are the wave group velocities in the x and y directions, and C_σ and C_θ are the propagation velocities in spectral space (σ, θ). S_{tot} is the sum of the source terms for wave variance, which can be defined as:

$$S_{tot} = S_{in} + S_{wc} + S_{nl4} + S_{nl3} + S_{br} + S_{bf}, \tag{11}$$

where S_{in} is the wind-induced energy input; S_{nl3} and S_{nl4} are the nonlinear interactions in shallow and deep-water areas, respectively; S_{wc} and S_{br} represent the wave energy dissipation in deep- and shallow-water areas caused by white capping and wave breaking, respectively; and S_{bf} is the dissipation caused by bottom friction. The first three source terms in Equation (11) are significant in deep water, while the last three terms are influential in shallow water. A peak enhancement of 3.3 and a bottom friction factor of 0.067 have been specified in WWM-III following a report from the Joint North Sea Wave Project (JONSWAP, Hasselmann et al. [29]). A wave-breaking coefficient of 0.78 is adopted for computing wave breaking in WWM-III. Nonlinear wave-wave interactions are treated with the discrete interaction approximation (DIA) proposed by Hasselmann et al. [30]. The wave direction is equally discretized into 36 bins over a range of 0–360°. The number of frequency bins is also 36, and the low- and high-frequency limits of the discrete wave period in WWM-III are 0.04 Hz and 1.0 Hz, respectively.

2.5. Fully Coupled SCHISM-WWM-III Model

SCHISM-2D and WWM-III take advantage of sharing the same subdomains and parallelization via the same domain decomposition scheme to eliminate interpolation errors and enhance the efficiency of information exchange between the ocean circulation and wind wave models. The computational performance of the coupled model, SCHISM-WWM-III, can be made more efficient through the use of different time steps in the two models; thus, time steps of 120 s and 600 s were assigned to the SCHISM and WWM-III models, respectively. This means that, each time information exchange is conducted between the two models, five time intervals of SCHISM-2D will have been calculated. The depth-averaged tidal current and tidal elevation are passed to WWM-III from SCHISM-2D, while WWM-III sends the radiation stresses to SCHISM-2D. The SCHISM-WWM-III modeling system has been successfully applied to hindcast, simulate, and predict storm surge and storm wave hazards and to evaluate the wave power energy in Taiwanese waters [2,3]. Further details about the coupling approaches used in SCHISM-WWM-III can be found in Reference [31].

2.6. Computational Domain and Unstructured Grid

Satisfying all requirements of the model domain and mesh is essential for successful typhoon wave and storm surge modeling. An appropriate computational domain size must be large enough to accommodate the entire typhoon while minimizing the effects of boundary conditions [32,33]. Additionally, the resolution of the model mesh must be high enough to resolve the geometric details of the surf zones and shallow-water regions [34–36]. Figure 2 shows the coverage of the model domain specified for application to Typhoons Soudelor and Megi. The full computational extent of the SCHISM-WWM-III computational domain is from 105° E to 140° E and from 15° N to 31° N, and it comprises 276,639 unstructured grid points and 540,510 nonoverlapping triangles. The mesh is coarser in the open ocean beyond the nearshore region, with a resolution of 20–40 km. For the coastal areas of Taiwan and its offshore islands, a finer mesh with a resolution of 200–400 m is employed (as shown in Figure 2).

The gridded bathymetric data used in SCHISM-WWM-III are composed of a global-scale dataset and a local-scale dataset. The GEBCO_2019 product released in March 2019 by the General Bathymetric Chart of the Oceans (GEBCO) provides global coverage on a 15 arc-second grid. The local-scale dataset, with a spatial resolution of 200 m, was acquired from the Department of Land Administration and the Ministry of the Interior, Taiwan, and covers the area from 100° E to 128° E and from 4° N to 29° N.

Figure 3 illustrates the seafloor elevations used in SCHISM-WWM-III, which were interpolated by combining the two datasets mentioned above.

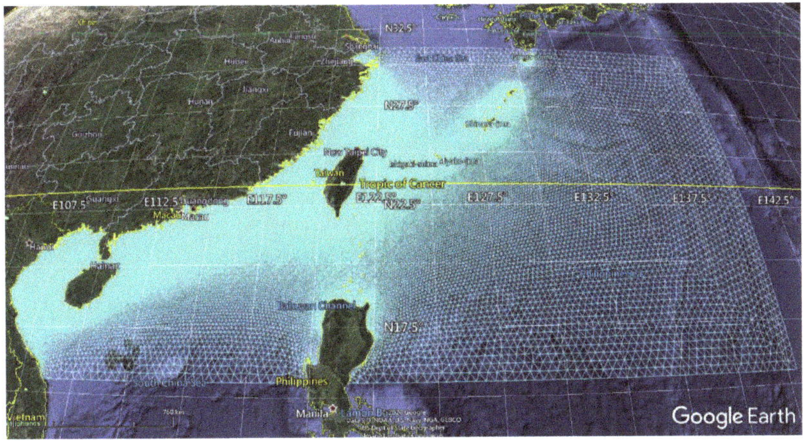

Figure 2. Coverage of the computational domain and its unstructured triangular mesh.

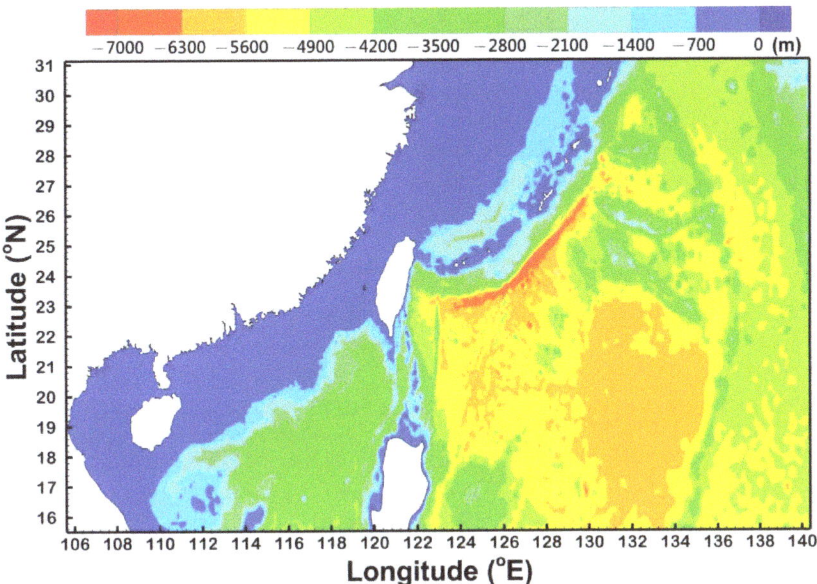

Figure 3. Distribution of seafloor elevations in the computational domain of Semi-implicit Cross-scale Hydroscience Integrated System Model Wind Wave Model version III (SCHISM-WWM-III).

2.7. Boundary Conditions

2.7.1. ERA5 Reanalysis Wind Forcing

ERA5 is the latest global climate atmospheric reanalysis released by the European Centre for Medium-Range Weather Forecasts (ECMWF). Compared to the ERA-Interim product, ERA5 has many improvements, e.g., much higher temporal and spatial resolutions, much improved tropospheric data, a better global balance between precipitation and evaporation, better precipitation over land in the deep tropics, and an improved representation of tropical cyclones, the last of which is the most important for simulating typhoon-generated waves. The hourly ERA5 atmospheric parameters, such as the wind at 10 m above the sea surface and the sea-level pressure, can be downloaded from the website of the Climate Data Store and have a spatial resolution of 30 km in the horizontal direction and 137 levels in the vertical direction. Quality-assured monthly updates of ERA5 are released within 3 months of real time. The ERA5 dataset extending back to 1950 is expected to become available in 2020.

2.7.2. Modified ERA5 Wind Forcing

Many previous studies have indicated that reanalysis products underestimate typhoon wind speeds, especially for the far-field regions of typhoons [37–42]. The peak typhoon wind speed can be considerably improved through a superposition method by combining parametric typhoon and reanalysis winds. Notably, a time mismatch sometimes exists between hindcast and measured winds. Although reanalysis products undervalue the high-speed winds of typhoons, they are more consistent in terms of the occurrence times of the peak winds. Therefore, a direct modification for reanalysis wind data (i.e., the ERA5 data in the present study) has been proposed by Pan et al. [39]. This modification is expressed as:

$$
W_{MERA5} = \begin{cases} W_{ERA5}\left[\frac{r}{R_{max}}\left(\frac{W_{Bmax}}{W_{Emax}}-1\right)+1\right] & r < R_{max} \\ W_{ERA5}\left[\frac{R_{trs}-r}{R_{trs}-R_{max}}\left(\frac{W_{Bmax}}{W_{Emax}}-1\right)+1\right] & R_{max} \leq r \leq 7R_{max} \\ W_{ERA5} & r > R_{trs} \end{cases} \tag{12}
$$

where W_{MERA5} is the modified wind speed at an arbitrary point in the computational domain, W_{ERA} is the wind speed from ERA5 at an arbitrary point in the computational domain, W_{Bmax} is the maximal wind speed of the typhoon as reported by the Regional Specialized Meteorological Center (RSMC) Tokyo–Typhoon Center, W_{Emax} is the maximal wind speed of the typhoon from ERA5, r is the radial distance from the center of the typhoon to an arbitrary point in the computational domain, and R_{max} is the radius of the maximal typhoon wind speed, R_{trs} is the radius of the transitional zone which was set to $6R_{max}$. R_{max} can be described as a function of W_{Bmax} and the latitude of the typhoon's center (ϕ):

$$
R_{max} = m_0 + m_1 + m_2(\phi - 25). \tag{13}
$$

Knaff et al. [43] utilized a standard variational technique to obtain the coefficients of Equation (13) for the Western Pacific typhoon basin. In accordance with their report, the values of m_0, m_1, and m_2 were set to 38.0 (in n mi), -0.1167 (in n mi kt^{-1}), and -0.0040 (in n mi $^{\circ-1}$), respectively, in the present study. The modified ERA5 (MERA5) winds were employed as another source of atmospheric conditions for comparison with those from ERA5 in terms of the performance of typhoon wave modeling.

2.7.3. Tidal Forcing

The harmonic parameters of amplitude and phase for the tidal elevation and horizontal velocity (u and v) were extracted from a regional inverse tidal model (China Seas & Indonesia [44]). M$_2$, S$_2$, N$_2$, K$_2$, K$_1$, O$_1$, P$_1$, and Q$_1$ serve as the eight main tidal constituents that drive SCHISM-WWM-III. Additionally, an inverse barometric effect on the boundary tidal elevation resulting from the difference between the actual atmospheric pressure and a reference atmospheric pressure (1013.0 hPa) was included in SCHISM-WWM-III. Specifying the open boundary forcing for the waves was not required since the wave-generating typhoons were fully contained within the model domain developed in the present study during the period of typhoon wave simulation. This approach is consistent with a previous study in which storm tides and waves were simulated using a large-scale modeling domain [45].

3. Results

The wind speeds at 10 m above the sea surface and the sea-level pressures during Typhoon Soudelor from 1–15 August in 2015 as extracted from ERA5 served as atmospheric forcing for SCHISM-WWM-III. The ERA5 winds were modified using the direct modification method expressed in Equation (12), and the typhoon winds derived from MERA5 were also imposed in SCHISM-WWM-III to compare the resulting storm wave simulation performance with that achieved using the ERA5 winds. The sensitivity of the simulation of the significant wave height during the typhoon period to the tidal elevation and tidal current was investigated by conducting several model experiments, as listed in Table 2.

Table 2. Overview of the designed model runs.

Scenario	Tidal Elevation	Tidal Current
S1	O	O
S2	O	X
S3	X	O
S4	X	X

O: included in wave simulation; X: excluded in wave simulation.

3.1. Effects of Atmospheric Forcing on Typhoon Wave Simulations

The hourly typhoon winds used in SCHISM-WWM-III were acquired from ERA5 at a horizontal resolution of 31 km and were then interpolated from the structured grid to the unstructured grid by means of the inverse distance weighting method. Figure 4a–c illustrate the instantaneous spatial distributions of the wind field during Typhoon Soudelor in 2015. As seen in Figure 4a–c, the wind speeds from the hourly ERA5 product reach only 24–36 m/s in the near-field region of Typhoon Soudelor. The instantaneous distributions of the simulated significant wave height corresponding to the hourly ERA5 winds at the specified times are displayed in Figure 4d–f. The simulated peak significant wave heights are 10–12 m in the deep-water area far from Taiwan (Figure 4d) and gradually increase to 12–13 m (Figure 4e) and 13–15 m (Figure 4f) with the typhoon's approach to Taiwan. Figure 5a–c present snapshots of the spatial distributions of the typhoon wind field obtained from MERA5 at the same specified times as in Figure 4a–c. Increases in wind speed are obvious near the inner region of the typhoon, and the maximal wind speed reaches 36–40 m/s or even exceeds 40 m/s after modification. Figure 5d–f show that the increases in the significant wave height caused by the use of the MERA5 winds relative to the values induced by the winds from ERA5 are 1–2 m. Additionally, the areas with large significant wave heights (greater than or equal to 15 m) are extended.

Comparisons of the significant wave height time series between model simulations using the different wind forcings and the corresponding measurements are presented in Figure 6a–c for the Fuguijiao (Figure 6a), Guishandao (Figure 6b), and Hualien (Figure 6c) buoys. The peak wave heights are underestimated by 2–2.5 m for all three wave buoys when the ERA5 winds are used in SCHISM-WWM-III. The underestimations between the simulated and measured peak wave heights are reduced to 0.5–1.0 m when the winds from MERA5 are imposed in SCHISM-WWM-III. Significant improvements in the simulated typhoon wave heights are achieved when employing the MERA5 winds; therefore, the SCHISM-WWM-III model combined with the winds from MERA5 was further utilized to investigate the sensitivity of storm wave simulations to the tidal elevation and tidal current for Typhoon Soudelor in 2015.

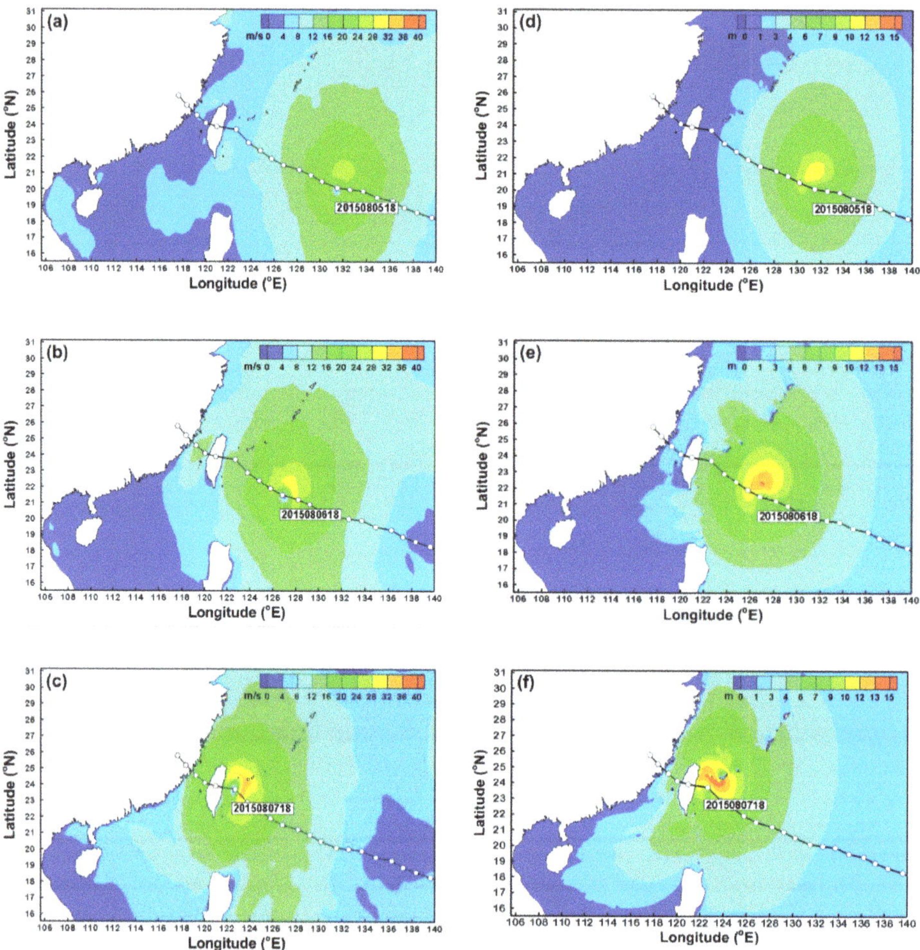

Figure 4. Instantaneous distributions of the hourly wind field from the fifth-generation global atmospheric reanalysis (ERA5) product (**a**–**c**) and the corresponding simulated significant wave heights (**d**–**f**) for Typhoon Soudelor at the specified times in 2015.

3.2. Typhoon Wave Simulation in the Absence of Tidal Current

The significant wave heights from 1–15 August in 2015 were reproduced using the MERA5-driven SCHISM-WWM-III model. Two scenarios were considered to better understand the effect of the tidal current on the wave modeling performance. One scenario was a wave simulation that included both the tidal elevation and tidal current (hereafter denoted by S1, as shown in Table 2), and the other was a wave simulation that included only the tidal elevation (hereafter denoted by S2, as shown in Table 2). Figure 7 shows comparisons of the significant wave height time series between the modeled and measured values for the Fuguijiao (Figure 7a), Guishandao (Figure 7b), and Hualien (Figure 7c) wave buoys using the winds from MERA5 in these two different scenarios. The S1 simulations show fluctuations in the significant wave height time series during the approach and retreat of the typhoon; by contrast, the S2 simulations, i.e., without the tidal current, are smoother. Notably, however, this phenomenon is only obvious at the Fuguijiao wave buoy (Figure 7a) and thus might be due to the presence of a stronger tidal current in the northern nearshore waters of Taiwan. The simulated peak significant wave heights from the S1 and S2 simulations are also depicted in a magnified form in Figure 7a–c. The peak significant wave height simulated in scenario S2 is slightly higher than that in S1 for the Fuguijiao wave buoy but is slightly lower for the Guishandao and Hualien wave buoys. Overall, the effect of the tidal current on the storm wave simulations during the period of Typhoon Soudelor in 2015 is most evident in the northern waters of Taiwan. The data issued from the industrial technology research institute of Taiwan implies that the averaged tidal current speed could be reached 0.8–1.0 m/s near the Fuguijiao wave buoy where the coastline has a cape shape, but only 0.2–0.3 m/s and 0.6–0.8 m/s in the nearshore waters adjacent to the Guishandao and Hualien wave buoys. For this reason, the typhoon wave simulations are more sensitivity to tidal current for the Fuguijiao buoy than for the others.

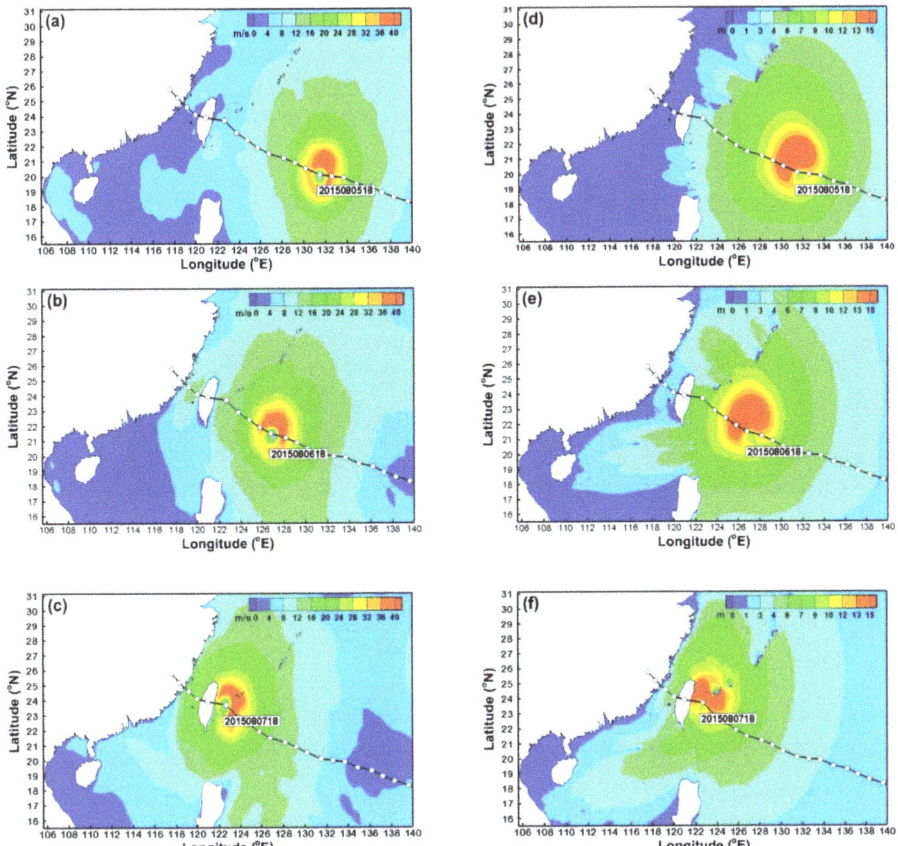

Figure 5. Instantaneous distributions of the hourly wind field from the modified fifth-generation global atmospheric reanalysis (MERA5) product (**a–c**) and the corresponding simulated significant wave heights (**d–f**) for Typhoon Soudelor at the specified times in 2015.

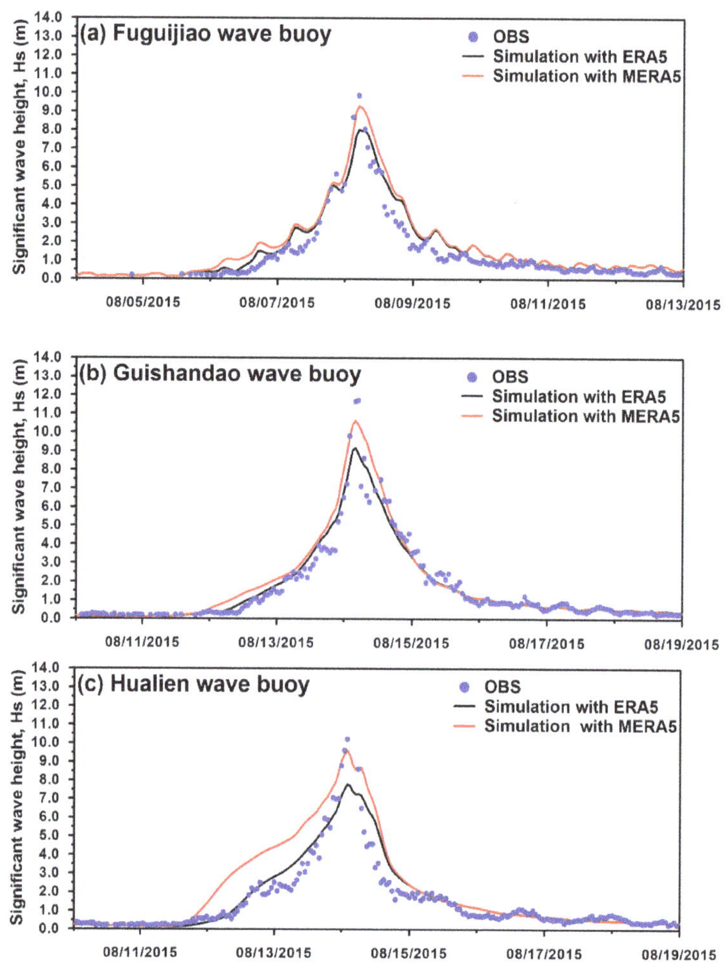

Figure 6. Comparison of the time series of the significant wave heights between simulations using different wind forcings and the corresponding measurements for the (**a**) Fuguijiao, (**b**) Guishandao, and (**c**) Hualien wave buoys during the period of Typhoon Soudelor in 2015.

3.3. Typhoon Wave Simulation in the Absence of Tidal Elevation

As shown in Table 2, the storm wave simulation scenario that includes the tidal current but not the tidal elevation is called S3. Comparisons of the significant wave heights between the S1 and S3 simulations and the corresponding measurements for the different wave buoys are illustrated in Figure 8, where Figure 8a–c present the results for the Fuguijiao, Guishandao, and Hualien wave buoys, respectively. Fluctuations are evident in Figure 8a (for the Fuguijiao wave buoy) for the scenario in which the tidal current is included for typhoon wave modeling. Interestingly, the fluctuation amplitudes are smaller than those in the S1 simulations. Additionally, phase shifts between the S1 and S3 simulations can also be seen for the Fuguijiao wave buoy (Figure 8a). For the Guishandao (Figure 8b) and Hualien (Figure 8c) wave buoys, the S1 and S3 simulations are almost identical with regard to the time series of the simulated significant wave heights. The differences in the simulated peak wave heights between the S1 and S3 simulations are also small, as seen from the enlarged views in Figure 8a–c. The observed phenomena may be attributable to the different tidal ranges in the northern, northeastern, and eastern coastal waters of Taiwan. According to the statistics of the long-term tidal monitoring data from the CWB, the average tidal ranges are 1.61, 0.88, and 1.0 m in the northern, northeastern, and eastern coastal waters of Taiwan, respectively. Thus, the tidal range near the Fuguijiao wave buoy (in the northern waters) is approximately twice as large as those near the Guishandao (northeastern) and Hualien (eastern) wave buoys; therefore, the effect of the tidal elevation on the wave simulations is more significant for the Fuguijiao wave buoy than for the Guishandao and Hualien wave buoys.

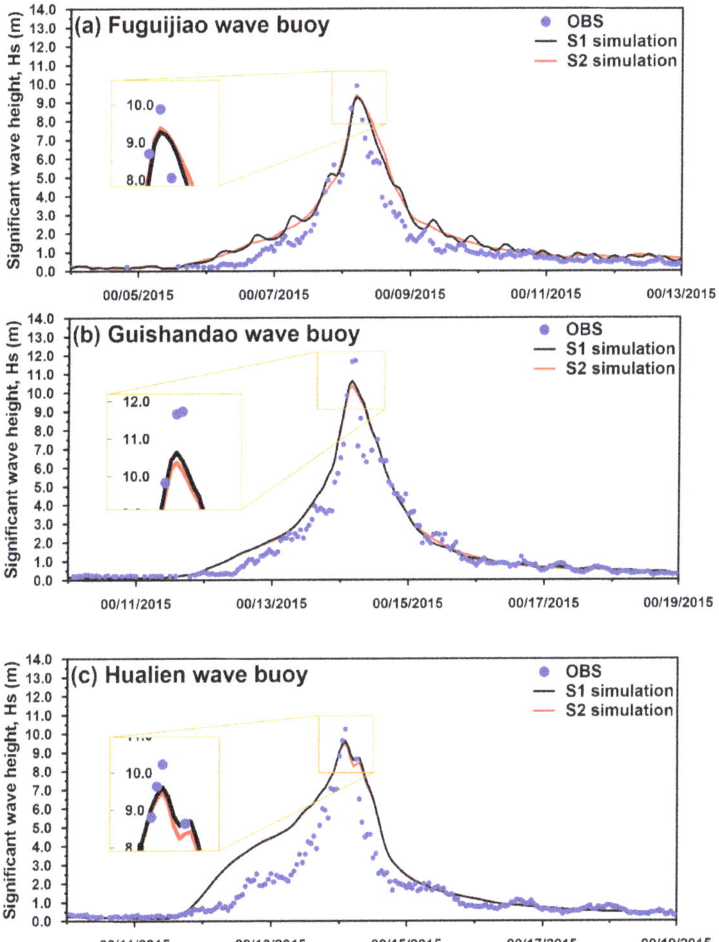

Figure 7. Comparison of the time series of the significant wave heights between the wave simulation that included both the tidal elevation and tidal current (S1) and the wave simulation that included only the tidal elevation (S2) and the corresponding measurements for the (**a**) Fuguijiao, (**b**) Guishandao, and (**c**) Hualien wave buoys during the period of Typhoon Soudelor in 2015.

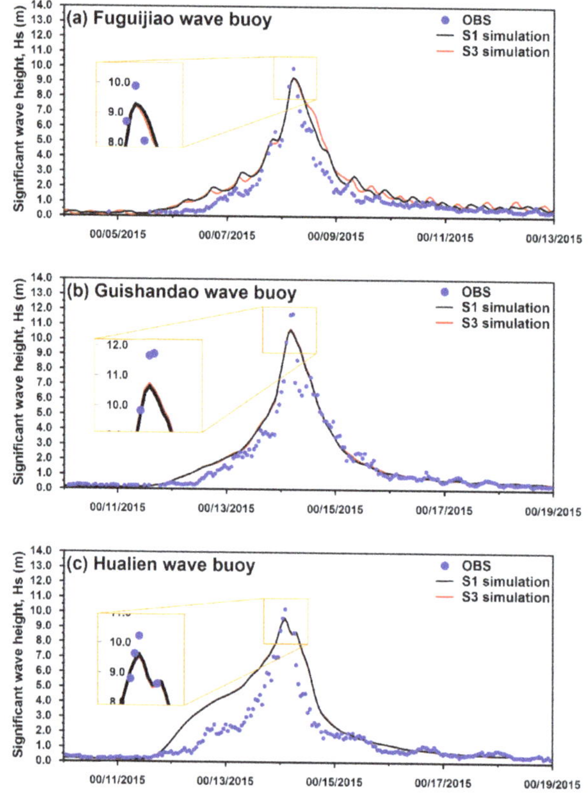

Figure 8. Comparison of the time series of the significant wave heights between the S1 and the storm wave simulation scenario that includes the tidal current but not the tidal elevation (S3) and the corresponding measurements for the (**a**) Fuguijiao, (**b**) Guishandao, and (**c**) Hualien wave buoys during the period of Typhoon Soudelor in 2015.

3.4. Typhoon Wave Simulation in the Absence of Both Tidal Elevation and Tidal Current

The S4 simulations can be regarded as stand-alone wave modeling because both tidal current and tidal elevation are neglected in the model (as shown in Table 2). Figure 9a–c present the model-data comparisons between the S1 and S3 simulations and the corresponding observations of the significant wave heights at the Fuguijiao (Figure 9a), Guishandao (Figure 9b), and Hualien (Figure 9c) wave buoys. As seen from Figure 9a, the significant wave height time series again becomes smooth when the tidal current is excluded from the typhoon wave simulation. Compared to the results in Figure 7a–c, no differences in the simulated significant wave height can be easily detected between the S2 (without the tidal current effect) and S4 simulations, even at the peak wave heights, for the three wave buoys. These findings reveal that the storm wave simulations are more sensitive to the absence of the tidal current than to the absence of the tidal elevation.

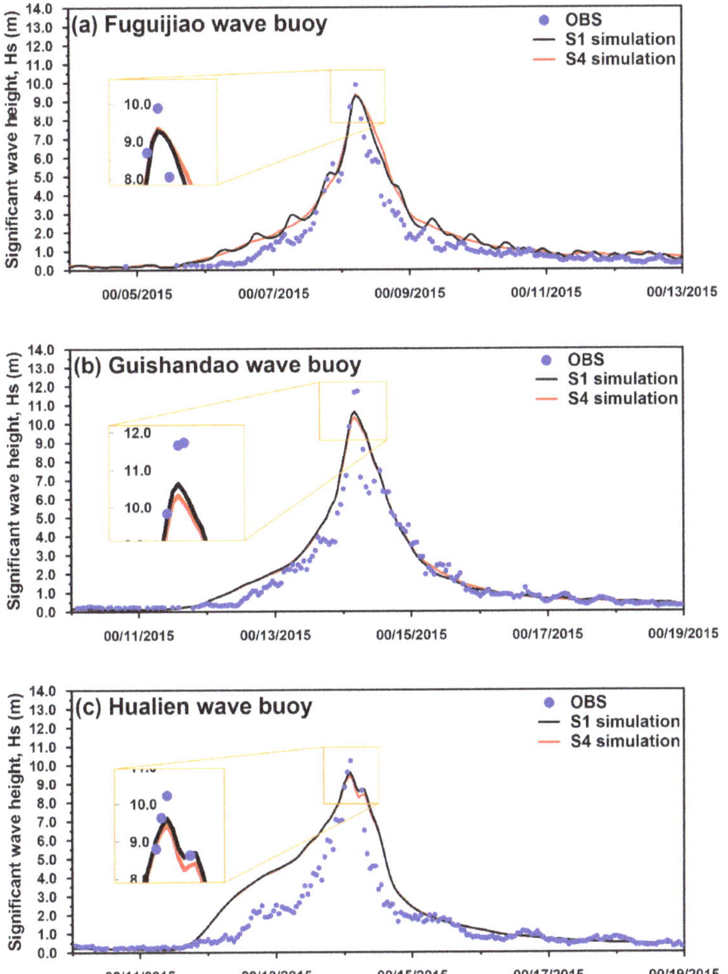

Figure 9. Comparison of the time series of the significant wave heights between the S1 and stand-alone wave modeling (S4) simulations and the corresponding measurements for the (**a**) Fuguijiao, (**b**) Guishandao, and (**c**) Hualien wave buoys during the period of Typhoon Soudelor in 2015.

4. Discussion

The four model experiments conducted in the present study indicate that the simulations of the significant wave heights during Typhoon Soudelor are more sensitive to the tidal current than to the tidal elevation. To understand the effects of the tidal current and tidal elevation on other wave parameters, the sensitivity of the mean wave period and wave direction to the tidal current and tidal elevation was also investigated for the Fuguijiao, Guishandao, and Hualien wave buoys during the period of Typhoon Soudelor in 2015. The simulated results for the mean wave periods and wave directions for the different wave buoys are compared in Figure 10. The greatest inconsistencies among the S1–S4 simulations for the mean wave period (Figure 10a–c) and wave direction (Figure 10d–f, where wave direction measurements for the Guishandao and Hualien wave buoys are unavailable) are observed for the Fuguijiao wave buoy (Figure 10a,b), and the simulations in scenarios S2 and S4 are almost identical. These findings are similar to those for the simulations of the significant wave height.

The results of the present study are different from those of Huang et al. [9] because the seabed slope off the eastern coast of Taiwan is very steep.

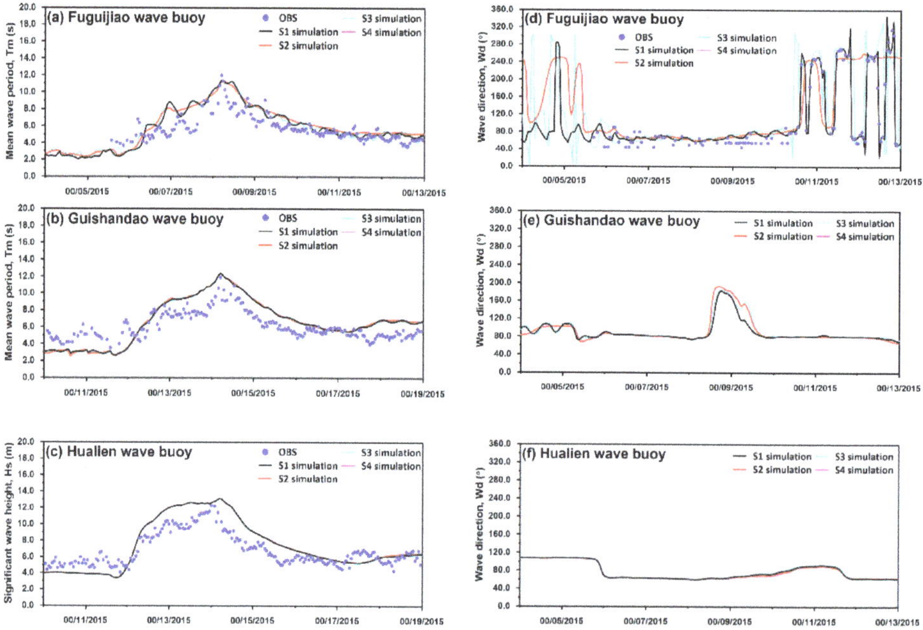

Figure 10. Comparison of the time series of the mean wave periods (**a–c**) and wave directions (**d–f**)between the S1, S2, S3, and S4 simulations and the corresponding measurements for the Fuguijiao (**a,d**), Guishandao (**b,e**), and Hualien (**c,f**) wave buoys during the period of Typhoon Soudelor in 2015.

A typhoon with a very similar track and intensity that struck Taiwan in 2016, namely, Typhoon Megi (as shown in Figure 1b), was adopted to confirm the reliability of the results derived in Section 3. Figure 11a–c compare the significant wave height time series between simulations using wind data from ERA5 and MERA5 and the corresponding measurements during the period of Typhoon Megi in 2016. Although the simulated peak significant wave heights are still underestimated for the Guishandao and Hualien wave buoys (Figure 11b,c), the MERA5-driven peak significant wave heights are increased by 2 m compared to those induced by the ERA5 winds. This observation indicates that the peak storm waves can be accurately simulated under the condition that the reanalysis typhoon winds are suitably modified. Numerical experiments analogous to those conducted in Section 3 for Typhoon Soudelor in 2015 were performed to investigate the sensitivity of the wave simulations to the tidal current and tidal elevation during the period of Typhoon Megi in 2016. The results for the simulated significant wave heights in scenarios S1–S4 are compared in Figures 12–14. Wave-like fluctuations are again observed in the significant wave height time series from the S1 and S2 simulations (i.e., wave modeling with and without the tidal current) for the Fuguijiao wave buoy (see the black lines in Figure 12a, Figure 13a, and Figure 14a and the red line in Figure 13a). However, this phenomenon is absent when the tidal current is excluded in the storm wave modeling (i.e., the S2 and S4 simulations; see the red lines in Figures 12a and 14a). Figure 15 presents the effects of the tidal current and tidal elevation on the simulated mean wave periods and wave directions for the different wave buoys during Typhoon Megi. The simulations for the Fuguijiao wave buoy show higher sensitivity to the tidal current (as shown in Figure 15a,d) than those for the Guishandao and Hualien wave buoys do (Figure 15b–f). Thus, it is again evident that the inclusion of the tidal current is more important for typhoon wave simulation for the Fuguijiao buoy than for the others.

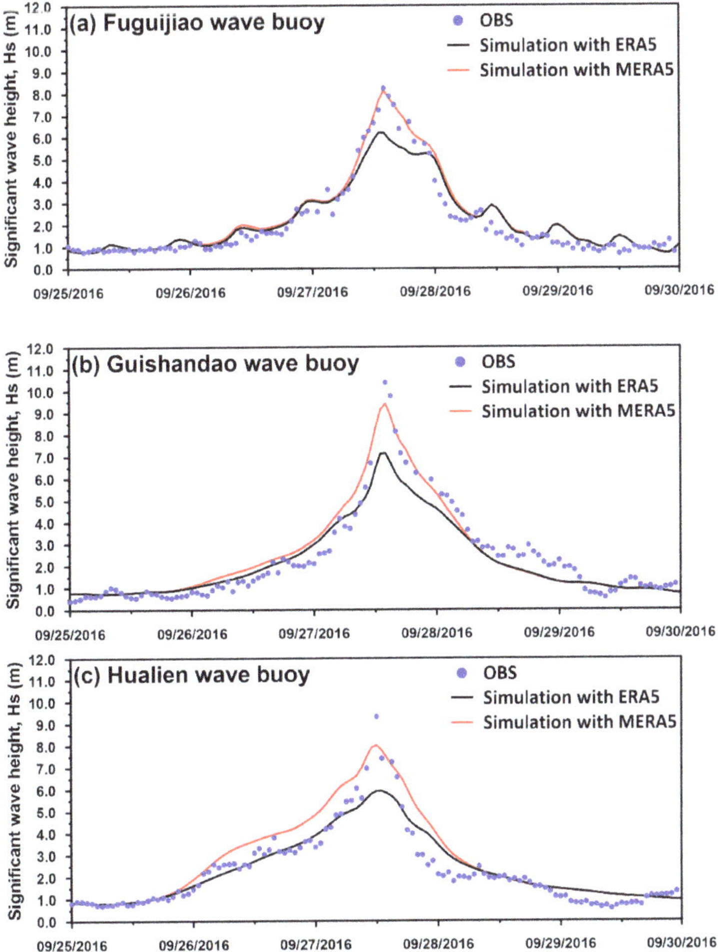

Figure 11. Comparison of the time series of the significant wave heights between simulations using different wind forcings and the corresponding measurements for the (**a**) Fuguijiao, (**b**) Guishandao, and (**c**) Hualien wave buoys during the period of Typhoon Megi in 2016.

Although the employment of the MERA5 typhoon winds improved the peak significant wave height simulations, a rise in the wave heights outside the peak was also evident, especially for the wave buoy at Hualien (as shown in Figures 6c and 11c). The adoption of the typhoon winds from the ERA5 gave better predictions of the significant wave height in the period before the peak. This phenomenon might be due to the overestimation of wind speed in the left side of typhoons by using the direct modification method. Thus, the wind speed comparisons between the ERA5, MERA5, and observation need further analysis in the future.

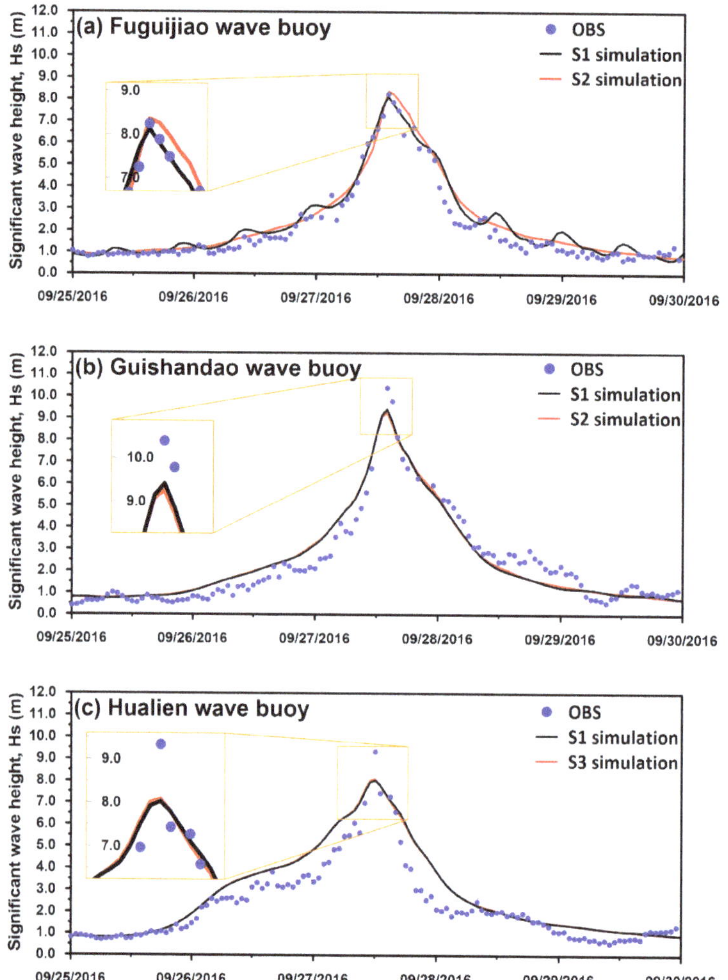

Figure 12. Comparison of the time series of the significant wave heights between the S1 and S2 simulations and the corresponding measurements for the (**a**) Fuguijiao, (**b**) Guishandao, and (**c**) Hualien wave buoys during the period of Typhoon Megi in 2016.

Saprykina et al. [46] indicated that the local wave field could be changed sharply, and the phenomenon is obvious in the presence of the adverse current. The discrepancy between the modeled and measured wave parameters in the present study could be due to the evolution of nonlinear modulation instability of waves and the increase in wave steepness on a gradually increasing adverse current.

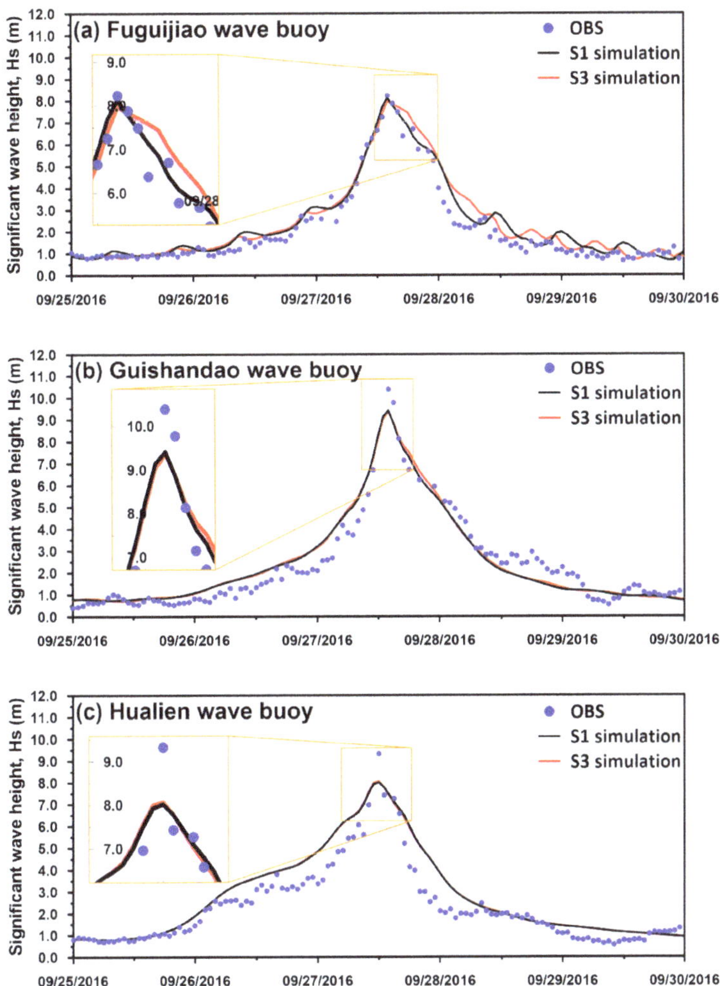

Figure 13. Comparison of the time series of the significant wave heights between the S1 and S3 simulations and the corresponding measurements for the (**a**) Fuguijiao, (**b**) Guishandao, and (**c**) Hualien wave buoys during the period of Typhoon Megi in 2016.

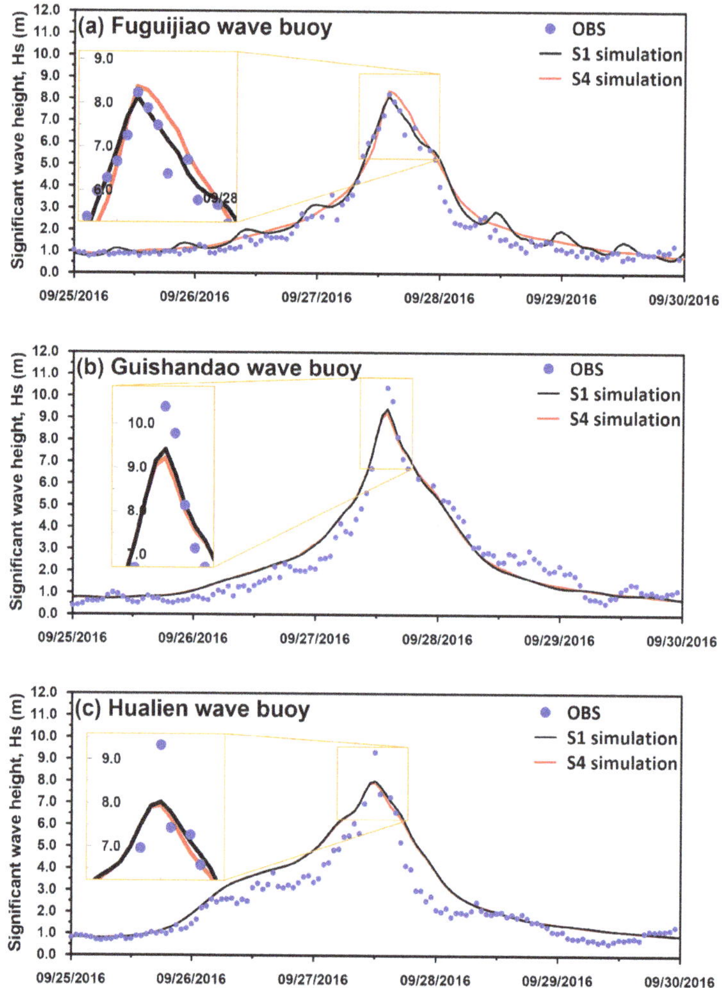

Figure 14. Comparison of the time series of the significant wave heights between the S1 and S4 simulations and the corresponding measurements for the (**a**) Fuguijiao, (**b**) Guishandao, and (**c**) Hualien wave buoys during the period of Typhoon Megi in 2016.

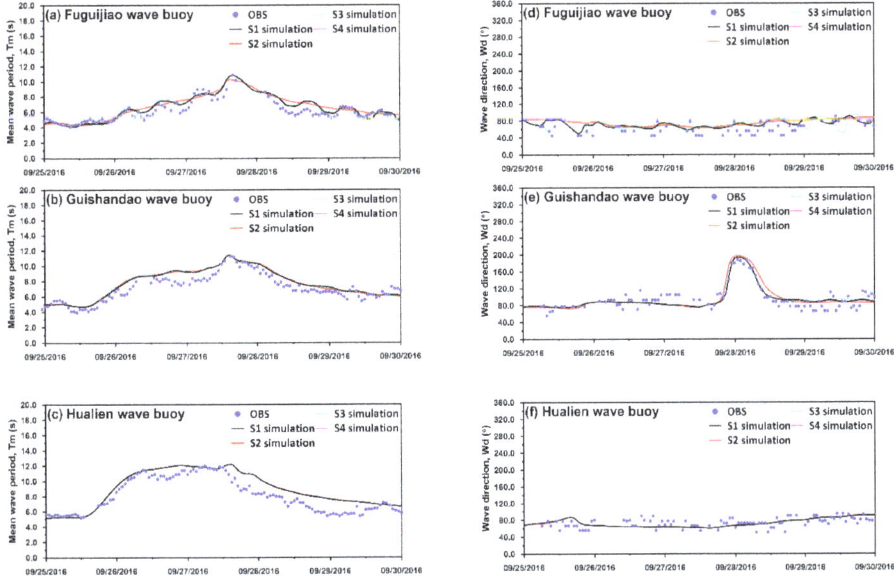

Figure 15. Comparison of the time series of the mean wave periods (**a–c**) and wave directions (**d–f**) between the S1, S2, S3, and S4 simulations and the corresponding measurements for the Fuguijiao (**a,d**), Guishandao (**b,e**), and Hualien (**c,f**) wave buoys during the period of Typhoon Megi in 2016.

5. Summary and Conclusions

This paper investigated the role of the tidal current and tidal elevation in wave simulations for the nearshore waters of Taiwan during typhoon events. A large computational domain with a high-resolution, unstructured-grid mesh was developed, and a fully coupled circulation-wave model, SCHISM-WWM-III, was applied to simulate the typhoon waves. The hourly typhoon winds from ERA5 and a modification of the original ERA5 product (MERA5) were considered as the atmospheric conditions for SCHISM-WWM-III to determine the optimal wind source for storm wave modeling. The performance achieved when adopting the MERA5 typhoon winds for significant wave height simulation was better than that achieved using the ERA5 winds. The simulated peak significant wave heights could be increased by 1.0–2.0 m through the utilization of the MERA5 winds for Typhoon Soudelor in 2015 and Typhoon Megi in 2016. A series of model experiments was then designed and conducted using SCHISM-WWM-III combined with the MERA5 wind data. The results of the analysis reveal that the tidal current is influential in the simulation of storm waves in the northern nearshore waters of Taiwan, where a stronger tidal current is present. The tidal current affects the simulation of not only the significant wave height but also the mean wave period and wave direction in the same nearshore area for both Typhoon Soudelor in 2015 and Typhoon Megi in 2016. Regarding the effect of the tidal elevation on the simulation of typhoon waves, the sea area in the northern waters of Taiwan is slightly sensitive to whether the tidal elevation is included in the model because the tidal range is larger in this area than in the northeastern and eastern waters. In summary, the inclusion of both the tidal current and tidal elevation is necessary for accurate sea state modeling in typhoon wave simulations, and this is particularly true for nearshore waters with a strong tidal current and a large tidal range. Another important finding is that, even with the imposition of fifth-generation reanalysis typhoon winds on the circulation-wave model, the significant wave heights induced by typhoons are still underestimated. The direct modification method applied in the present study is an alternative approach for improving the simulation of the peak wave heights while maintaining a reliable structure of the entire typhoon wind field. In future research on the sensitivity of storm wave modeling, it will be important to investigate a wider variety of typhoons with different tracks.

Author Contributions: Conceptualization, S.-C.H., H.-L.W., W.-B.C., C.-H.C., and L.-Y.L.; methodology, W.-B.C. and H.-L.W.; validation, W.-B.C.; formal analysis, H.-L.W. and C.-H.C.; writing—original draft preparation, W.-B.C., S.-C.H., H.-L.W., and L.-Y.L.; supervision, S.C.H. and L.-Y.L. All authors have read and agreed to the published version of the manuscript.

Funding: This research was funded by the Ministry of Science and Technology (MOST), Taiwan, grant No. MOST 109-2221-E-865-001.

Acknowledgments: The authors would like to thank the Central Weather Bureau and Department of Land Administration, Ministry of the Interior of Taiwan for providing the measured data, as well as Joseph Zhang at the Virginia Institute of Marine Science, College of William & Mary, and Aron Roland at the BGS IT&E GmbH, Darmstadt, Germany, for kindly sharing their experiences concerning the use of the numerical model.

Conflicts of Interest: The authors declare no conflict of interest.

References

1. Shao, Z.; Liang, B.; Li, H.; Wu, G.; Wu, Z. Blended wind fields for wave modeling of tropical cyclones in the South China Sea and East China Sea. *Appl. Ocean Res.* **2018**, *71*, 20–33. [CrossRef]

2. Shih, H.-J.; Chang, C.-H.; Chen, W.-B.; Lin, L.-Y. Identifying the Optimal Offshore Areas for Wave Energy Converter Deployments in Taiwanese Waters Based on 12-Year Model Hindcasts. *Energies* **2018**, *11*, 499. [CrossRef]

3. Chang, C.-H.; Shih, H.-J.; Chen, W.-B.; Su, W.-R.; Lin, L.-Y.; Yu, Y.-C.; Jang, J.-H. Hazard Assessment of Typhoon-Driven Storm Waves in the Nearshore Waters of Taiwan. *Water* **2018**, *10*, 926. [CrossRef]

4. Lewis, M.; Bates, P.; Horsburgh, K.; Neal, J.; Schumann, G. A storm surge inundation model of the northern Bay of Bengal using publicly available data. *Q. J. R. Meteorol. Soc.* **2013**, *139*, 358–369. [CrossRef]

5. Wang, T.; Yang, Z.; Wu, W.-C.; Grear, M. A Sensitivity Analysis of the Wind Forcing Effect on the Accuracy of Large-Wave Hindcasting. *J. Mar. Sci. Eng.* **2018**, *3*, 139. [CrossRef]

6. Longuet-Higgins, M.S.; Stewart, R.W. Radiation stress in water waves: A physical discussion with applications. *Deep Sea Res.* **1964**, *11*, 529–562.

7. Phillips, O.M. *The Dynamics of Upper Ocean*; Cambridge University Press: New York, NY, USA, 1977; 336p.

8. Large, W.G.; Pond, S. Open ocean momentum fluxes in moderate to strong winds. *J. Phys. Oceanogr.* **1981**, *11*, 324–336. [CrossRef]

9. Huang, Y.; Weisberg, R.H.; Zheng, L. Coupling of surge and waves for an Ivan-like hurricane impacting the Tampa Bay, Florida region. *J. Geophys. Res.* **2010**, *115*, C12009. [CrossRef]

10. Hsiao, S.-C.; Chen, H.; Chen, W.-B.; Chang, C.-H.; Lin, L.-Y. Quantifying the contribution of nonlinear interactions to storm tide simulations during a super typhoon event. *Ocean Eng.* **2019**, *194*, 106661. [CrossRef]

11. Rusu, L.; Bernardino, M.; Guedes Soares, C. Influence of Wind Resolution on Prediction of Waves Generated in an Estuary. *J. Coast. Res.* **2009**, *56*, 1419–1423.

12. Van Vledder, G.P.; Akpınar, A. Wave model predictions in the Black Sea: Sensitivity to wind fields. *Appl. Ocean Res.* **2015**, *53*, 161–178. [CrossRef]

13. Zhang, Y.J.; Ye, F.; Stanev, E.V.; Grashorn, S. Seamless cross-scale modelling with SCHISM. *Ocean Modell.* **2016**, *102*, 64–81. [CrossRef]

14. Zhang, Y.J.; Baptista, A.M. SELFE: A semi-implicit Eulerian-Lagrangian finite-element model for cross-scale ocean circulation. *Ocean Modell.* **2008**, *21*, 71–96. [CrossRef]

15. Wang, H.V.; Loftis, J.D.; Liu, Z.; Forrest, D.; Zhang, J. The storm surge and sub-grid inundation modeling in New York City during hurricane Sandy. *J. Mar. Sci. Eng.* **2014**, *2*, 226–246. [CrossRef]

16. Chen, W.-B.; Liu, W.-C. Modeling flood inundation Induced by river flow and storm surges over a river basin. *Water* **2014**, *6*, 3182–3199. [CrossRef]

17. Chen, W.-B.; Liu, W.-C. Assessment of storm surge inundation and potential hazard maps for the southern coast of Taiwan. *Nat. Hazards* **2016**, *82*, 591–616. [CrossRef]

18. Chen, Y.-M.; Liu, C.-H.; Shih, H.-J.; Chang, C.-H.; Chen, W.-B.; Yu, Y.-C.; Su, W.-R.; Lin, L.-Y. An Operational Forecasting System for Flash Floods in Mountainous Areas in Taiwan. *Water* **2019**, *11*, 2100. [CrossRef]

19. Chen, W.-B.; Chen, H.; Lin, L.-Y.; Yu, Y.-C. Tidal Current Power Resource and Influence of Sea-Level Rise in the Coastal Waters of Kinmen Island, Taiwan. *Energies* **2017**, *10*, 652. [CrossRef]

20. Chen, W.-B.; Liu, W.-C. Investigating the fate and transport of fecal coliform contamination in a tidal estuarine system using a three-dimensional model. *Mar. Pollut. Bull.* **2017**, *116*, 365–384. [CrossRef]

21. Powell, M.D.; Vickery, P.J.; Reinhold, T.A. Reduced drag coefficient for high wind speeds in tropical cyclones. *Nature* **2003**, *422*, 279–283. [CrossRef]

22. Smith, S.D. Wind stress and heat flux over the ocean in gale force winds. *J. Phys. Oceanogr.* **1980**, *10*, 709–726. [CrossRef]

23. Longuet-Higgins, M.S.; Stewart, R.W. Radiation stress and transport in gravity wave, with application to 'surf beat'. *J. Fluid Mech.* **1962**, *13*, 485–504. [CrossRef]

24. Komen, G.J.; Cavaleri, L.; Donelan, M.; Hasselmann, K.; Hasselmann, S.; Janssen, P.A.E.M. *Dynamics and Modelling of Ocean Waves*; Cambridge University Press: Cambridge, UK, 1994; p. 532.

25. Roland, A. Development of WWM II: Spectral Wave Modeling on Unstructured Meshes. Ph.D. Thesis, Institute of Hydraulic and Water Resources Engineering, Technische Universität Darmstadt, Darmstadt, Germany, 2009.

26. Hsu, T.-W.; Ou, S.-H.; Liau, J.-M. Hindcasting nearshore wind waves using a FEM code for SWAN. *Coastal Eng.* **2005**, *52*, 177–195. [CrossRef]

27. Booij, N.; Ris, R.C.; Holthuijsen, L.H. A third-generation wave model for coastal regions. 1. Model description and validation. *J. Geophys. Res.* **1999**, *104*, 7649–7666. [CrossRef]

28. Su, W.-R.; Chen, H.; Chen, W.-B.; Chang, C.-H.; Lin, L.-Y.; Jang, J.-H.; Yu, Y.-C. Numerical investigation of wave energy resources and hotspots in the surrounding waters of Taiwan. *Renew. Energy* **2018**, *118*, 814–824. [CrossRef]

29. Hasselmann, K.; Barnett, T.P.; Bouws, E.; Carlson, H.; Cartwright, D.E.; Enke, K. *Measurements of Wind-Wave Growth and Swell Decay during the Joint North Sea Wave Project (JONSWAP)*; Deutsche Hydrographische Institut (DHI): Hamburg, Germany, 1973; Volume 12.

30. Hasselmann, S.; Hasselmann, K.; Allender, J.H.; Barnett, T.P. Computations and parameterizations of the non-linear energy transfer in a gravity-wave spectrum, Part 2: Parameterizations of the non-linear energy transfer for application in wave models. *J. Phys. Oceanogr.* **1985**, *15*, 1378–1391. [CrossRef]

31. Roland, A.; Zhang, Y.J.; Wang, H.V.; Meng, Y.; Teng, Y.-C.; Maderich, V.; Brovchenko, I.; Dutour-Sikiric, M.; Zanke, U. A fully coupled 3D wave-current interaction model on unstructured grids. *J. Geophys. Res.* **2012**, *117*, C00J33. [CrossRef]

32. Orton, P.; Georgas, N.; Blumberg, A.; Pullen, J. Detailed modeling of recent severe storm tides in estuaries of the New York City region. *J. Geophys. Res.* **2012**, *117*, C09030. [CrossRef]

33. Zheng, L.; Weisberg, R.H.; Huang, Y.; Luettich, R.A.; Westerink, J.J.; Kerr, P.C.; Donahue, A.S.; Grane, G.; Akli, L. Implications from the comparisons between two- and three-dimensional model simulations of the Hurricane Ike storm surge. *J. Geophys. Res. Ocean.* **2013**, *118*, 3350–3369. [CrossRef]

34. Zhang, H.; Sheng, J. Estimation of extreme sea levels over the eastern continental shelf of North America. *J. Geophys. Res. Ocean.* **2013**, *118*, 6253–6273. [CrossRef]

35. Muis, S.; Verlaan, M.; Winsemius, H.C.; Aerts, J.C.J.H.; Ward, P.J. A global reanalysis of storm surges and extreme sea levels. *Nat. Commun.* **2016**, *7*, 11969. [CrossRef] [PubMed]

36. Marsooli, R.; Lin, N. Numerical modeling of historical storm tides and waves and their interactions along the U.S. East and Gulf Coasts. *J. Geophys. Res. Ocean.* **2018**, *123*, 3844–3874. [CrossRef]

37. Chen, W.-B.; Chen, H.; Hsiao, S.-C.; Chang, C.-H.; Lin, L.-Y. Wind forcing effect on hindcasting of typhoon-driven extreme waves. *Ocean Eng.* **2019**, *188*, 106260. [CrossRef]

38. Murty, P.N.L.; Srinivas, K.S.; Rao Rama Pattabhi, E.; Bhaskaran, P.K.; Shenoi, S.S.C.; Padmanabham, J. Improved cyclone wind fields over the Bay of Bengal and their application in storm surge and wave computations. *Appl. Ocean Res.* **2020**, *95*, 102048. [CrossRef]

39. Pan, Y.; Chen, Y.P.; Li, J.X.; Ding, X.L. Improvement of wind field hindcasts for tropical cyclones. *Water Sci. Eng.* **2016**, *9*, 58–66. [CrossRef]

40. Li, J.; Pan, S.; Chen, Y.; Fan, Y.M.; Pan, Y. Numerical estimation of extreme waves and surges over the northwest Pacific Ocean. *Ocean Eng.* **2018**, *153*, 225–241. [CrossRef]

41. Qiao, W.; Song, J.; He, H.; Li, F. Application of different wind filed models and wave boundary layer model to typhoon waves numerical simulation in WAVEWATCH III model. *Tellus A* **2019**, *7*, 1657552. [CrossRef]

42. Hsiao, S.-C.; Chen, H.; Wu, H.-L.; Chen, W.-B.; Chang, C.-H.; Guo, W.-D.; Chen, Y.-M.; Lin, L.-Y. Numerical Simulation of Large Wave Heights from Super Typhoon Nepartak (2016) in the Eastern Waters of Taiwan. *J. Mar. Sci. Eng.* **2020**, *8*, 217. [CrossRef]

43. Knaff, J.A.; Sampson, C.R.; Demaria, M.; Marchok, T.P.; Gross, J.M.; Mcadie, C.J. Statistical Tropical Cyclone Wind Radii Prediction Using Climatology and Persistence. *Weather Forecast.* **2007**, *22*, 781–791. [CrossRef]

44. Zu, T.; Gana, J.; Erofeevac, S.Y. Numerical study of the tide and tidal dynamics in the South China Sea. *Deep Sea Res. Part I* **2008**, *55*, 137–154. [CrossRef]

45. Liu, Z.; Wang, H.; Zhang, Y.J.; Magnusson, L.; Loftis, J.D. Cross-scale modeling of storm surge, tide and inundation in Mid-Atlantic Bight and New York City during Hurricane Sandy 2012. *Estuar. Coast. Shelf Sci.* **2020**, *233*, 106544. [CrossRef]

46. Saprykina, Y.V.; Kuznetsov, S.Y.; Shugan, I.V.; Hwung, H.H.; Hsu, W.Y.; Yang, R.Y. Discrete evolution of the surface wave spectrum on a nonuniform adverse current. *Dokl. Earth Sci.* **2015**, *464*, 1075–1079. [CrossRef]

Journal of
Marine Science and Engineering

Article

Wave Simulation by the SWAN Model and FVCOM Considering the Sea-Water Level around the Zhoushan Islands

Zhehao Yang [1], Weizeng Shao [1,2,*], Yang Ding [3], Jian Shi [4] and Qiyan Ji [1]

[1] Marine Science and Technology College, Zhejiang Ocean University, Zhoushan 316022, China; yzh19960502@outlook.com (Y.Z.); jiqiyan@zjou.edu.cn (Q.J.)

[2] National Satellite Ocean Application Service, Beijing 100081, China

[3] Frontier Science Center for Deep Ocean Multispheres and Earth System (FDOMES) and Physical Oceanography Laboratory, Ocean University of China and Qingdao National Laboratory for Marine Science and Technology, Qingdao 266100, China; dingyangpol@ouc.edu.cn

[4] College of Meteorology and Oceanography, National University of Defense Technology, Nanjing 211101, China; shijian.mil@163.com

* Correspondence: shaoweizeng@mail.tsinghua.edu.cn; Tel.: +86-580-2550-753

Received: 1 September 2020; Accepted: 5 October 2020; Published: 8 October 2020

Abstract: In this study, the numerical wave model Simulating Waves Nearshore (SWAN), which resolves nearshore wave processes, and a hydrodynamic model, the Finite-Volume Community Ocean Model (FVCOM), were coupled to simulate waves and currents during Typhoon Fung-wong (2014) and Typhoon Chan-hom (2015) around the Zhoushan Islands. Both of these models employ the same unique unstructured grid. In particular, the influence of sea-surface currents, e.g., typhoon-induced and tidal currents, as well as the sea-water level, on wave simulation was studied. The composite wind field, which is derived from the parametric Holland model and European Centre for Medium-Range Weather Forecasts (ECMWF) winds (H-E winds), was taken as the forcing field. TPXO.5 tide data, sea-surface temperatures from the HYbrid Coordinate Ocean Model (HYCOM), HYCOM sea-surface salinity, and HYCOM sea-surface current were treated as open-boundary conditions. The comparison of sea-surface-current speed between the FVCOM simulation and the National Centers for Environmental Prediction (NCEP) Climate Forecast System Version 2 (CFSv2) data revealed a root-mean-square error (RMSE) of about 0.1206 m/s, with a correlation (Cor) more than 0.8, while the RMSE of the simulated sea-water level when compared with the HYCOM data was around 0.13 m, with a Cor of about 0.86. The validation indicated that the simulated results in this study were reliable. A sensitive experiment revealed that the sea-water level affected the typhoon-induced wave simulation. Validation against the measurements from the moored buoys showed an RMSE of <0.9 m for the sea-water level, which specifically reflected less overestimation during the high-sea state. Moreover, the significant-wave-height (SWH) difference (SWH without the sea-water level minus SWH with the sea-water level) was as great as −0.5 m around the Zhoushan Islands during the low-sea state. Furthermore, we studied the typhoon-induced waves when Typhoon Fung-wong passed the Zhoushan Islands, revealing that the reduction of SWH could be up to 1 m in the Yangtze Estuary and tidal flats when the maximum waves occurred.

Keywords: typhoon wave; SWAN; FVCOM; current; sea-water level

---

1. Introduction

Sea surface waves play an important role in the air–sea interaction layer. In particular, tropical cyclones induce extreme waves, which are a type of destructive nearshore disaster affecting human activity and offshore oil platform facilities. Given this, wave simulation is a preeminent topic for the oceanography community, especially for coastal waters [1,2] and tropical cyclones [3,4]. Two types of third-generation numerical wave models are popular for wave simulation, i.e., WAVEWATCH-III (WW3) [5], developed by the National Oceanic and Atmospheric Administration/National Centers for Environmental Prediction (NOAA/NCEP), and Simulating Waves Nearshore (SWAN) [6], developed by the Delft University of Technology. Waves are often affected by mesoscale/large-scale circulations, e.g., currents can shift wave energy due to the Doppler effect, and sea-water levels have an impact on wave propagation and dissipation due to depth-limited wave breaking. In these circumstances, the wave–current interaction is highly complex, especially under extreme weather conditions.

It is well known that the SWAN model performs well in the nearshore region, while the WW3 model is widely used for simulating waves in the open sea [7,8]. The advantage of the SWAN model is the application of an unstructured triangular grid employing an analog of the Gauss–Seidel sweeping method, which can portray complex coastal regions and numerous barriers, especially those of islands [9,10]. The discrete implementation of the SWAN model requires special terms, e.g., the nonlinear interactions [11] due to changes in bathymetry, wave-current, and white-cap breaking, which affect the evolution of wind-induced waves. The SWAN model utilized in this study (version 41.20, released in May 2019) has been successfully applied to coastal seas, estuaries [12], reefs, and lakes [13]. Tropical cyclones exhibit complicated physical dynamics at various spatial scales during their movement [14–16]. Strong wind-induced waves are generated in deep water, and storm surges on the continental shelf are large, both of which propagate and are transformed in the complex nearshore environment due to rapid changes in bathymetry and bottom friction. Wave dissipation also spreads over long time periods and large spatial scales and smoothly varies on continental shelves, which should be considered, particularly near islands or other breaking zones. Storm surges raise sea-water levels and then interact with waves and currents [17]. These issues clearly influence wave simulations, especially for tropical cyclones [18]. In the typhoon-modeling applications around the Zhoushan Islands for depths ranging from tens of meters to ten kilometers in the East China Sea, the dynamics at various spatial scales necessitate the implementation of unstructured grids, ensuring a variable water depth from kilometers in the deeper East China Seas to hundreds of meters on the continental shelf and tens of meters near the coastal shorelines.

The Finite-Volume Community Ocean Model (FVCOM) [19,20] is a coastal ocean circulation model developed by the University of Massachusetts and the Woods Hole Oceanographic Institution [20]. Its basic features include (1) an unstructured grid, (2) a finite volume, (3) free surfaces, and (4) 3D primitive equations. Similar to the SWAN model, the FVCOM uses the unstructured grid approach, which is advantageous for resolving dynamics in regions with complex irregular geometry, while the finite-element/finite-difference methods provide good numerical representations of the equations for momentum, mass, salt, heat, and tracer conservation. Moreover, the FVCOM makes use of improved computational fluid dynamics in order to better illustrate the convergence of numerical solutions for flow discontinuities and large gradients [21]. A comparison between the FVCOM and the two finite-difference ocean models, i.e., the Princeton Ocean Model (POM) and the Estuarine and Coastal Ocean Model (ECOM), demonstrated that numerical accuracy is significantly improved by depicting the physics of waves, tides, and flow in the coastal oceans [22]. The FVCOM has also been employed to examine the oceanic response to hurricane activity [23–25], and experiments have revealed that it accurately illustrates tide- and wind-induced rises in the water level [26].

In our previous research [26,27], the performance of input/dissipation source-term packages in the WAVEWATCH-III model was investigated in typhoons using rectangular meshes. Specifically, it was found that currents can affect typhoon-induced wave simulations at current speeds > 0.5 m/s and have a significant influence at current speeds > 2 m/s [28]. Experiments concerning the influence of the sea-water level on wave simulation based on the FVCOM and SWAN have already been performed [29]. It has been demonstrated that it is critical to include tide and sea-water-level variations in shelf and nearshore wave simulations [30,31]. The sea water level could be significantly affected by strong tide and typhoon-induced wind in complex coastal seas and then modulate the wave properties. Therefore, in this study, the SWAN and FVCOM, with unstructured grids, were coupled in order to simulate the waves in two typhoons passing the Zhoushan Islands in the East China Sea. A sensitive experiment was also performed to analyze the impact of sea-water levels on the wave simulations.

The remainder of this manuscript is organized as follows. The dataset is introduced in Section 2, i.e., the selected typhoons, forcing fields, and open boundary data, as well as the ancillary data used for validation. The model settings for the SWAN and FVCOM simulations are also briefly described. The accuracy of the simulated sea-surface-current speeds, sea-water levels, and ocean waves is confirmed in Section 3, a discussion is presented in Section 4, and the conclusions are summarized in Section 5.

2. Dataset Collection and Methods

Two typhoons, Fung-wong and Chan-hom, which passed over the Zhoushan Islands from 17 to 25 September 2014 and 29 June to 12 July 2015, respectively, were selected. The data for the two cyclones were collected from the Tokyo-Typhoon Center of the Japan Meteorological Agency (JMA). The tracks of the two typhoons overlaid on the water-depth map are presented in Figure 1a, in which the black square encloses the waters around the Zhoushan Islands. It can be seen in Figure 1b that the water depth ranges from kilometers in the East China Seas to tens of meters at the coastal shorelines. In particular, there is a series of islands at the mouth of Hangzhou Bay, indicating that the buoys located between Hangzhou Bay and the island chain provide valuable data, since the strong tidal currents, typhoon-induced currents, and changes in sea-water levels all occur in this area. This confluence of factors provides us with the opportunity to investigate the effects of typhoon-induced current terms and storm water terms on wave simulation for typhoons.

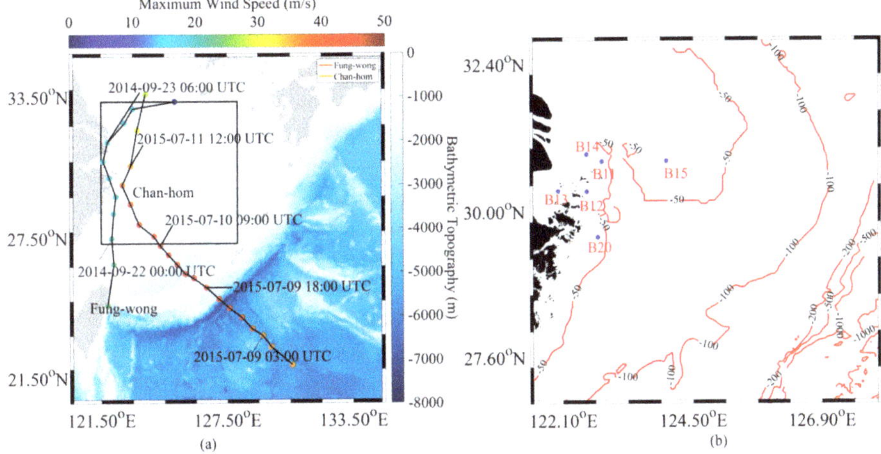

Figure 1. (**a**) Water-depth map of the selected region in the East China Sea, in which the red and blue lines represent the tracks of Typhoons Fung-wong (2014) and Chan-hom (2015), respectively, and the colored dots represent the maximum wind speeds of the typhoons. The area inside the black square is the geographic region around the Zhoushan Islands, China; (**b**) bathymetric topography of the study area corresponding to the black square in Figure 1a, in which the black dots are the available measurements from moored buoys.

Since 1979, the European Centre for Medium-Range Weather Forecasts (ECMWF) has helpfully provided atmospheric-marine data with a fine spatial resolution up to 0.125° (~12.5 km) at six-hour intervals every day. As for the analysis of global wave distributions, the ECMWF wind data are commonly used in the SWAN [32,33] and WW3 models [34–36]. In this study, however, a composite wind field, which was derived from the parametric Holland model and ECMWF winds (H-E winds), was taken as the forcing field for the model simulations. Specifically, the parametric Holland model was trained by fitting the shape parameter to buoy-measured observations (spots B_s in Figure 1b). The details of the H-E winds are discussed in Section 3 of [26], and the validation indicates a root-mean-square error (RMSE) less than 3 m/s for the shape parameter treated as a constant (0.4). Moreover, currents were also included in the wave simulation using the SWAN model [28]. The simulated waves in the East China Sea lasted from 1 September to 1 October 2014 and from 1 July to 1 August 2015. The modeling region was located between 20° and 35°N latitude and 120° and 135°E longitude, in which the bathymetric topography was provided by the General Bathymetric Chart of the Oceans (GEBCO) with a 0.01° (~1 km in the horizontal direction) gridded spatial resolution.

It is well known that storm water inundation is an important factor for assessing marine-hazard risks under extreme weather conditions [37] and also impacts wave simulation for coastal waters through various effects, e.g., wave-induced radiation stress, current–wave energy exchange, and bottom stress. Therefore, the sea-surface current and sea-water level were simulated with the FVCOM for the same region as the SWAN model, in which the H-E wind was the forcing field. The tide data from the TPXO.5 model, as well as the sea-surface temperature, sea-surface salinity, and sea-surface current values derived at 60-minute intervals at a 1/12° (~8 km) grid spatial resolution from the HYbrid Coordinate Ocean Model (HYCOM) were used for the open-boundary condition. The unstructured grids are illustrated in Figure 2 and are uniquely used in the SWAN and FVCOM simulations. It should be noted that the simulated currents lasted from 1 July to 1 October 2014 and from 1 May to 1 August 2015, to ensure the stability of the FVCOM [38]. The basic governing equations of the SWAN and FVCOM simulations are described in Appendices A and B, respectively. The details concerning the model settings for the FVCOM and SWAN are briefly presented in Tables 1 and 2, respectively.

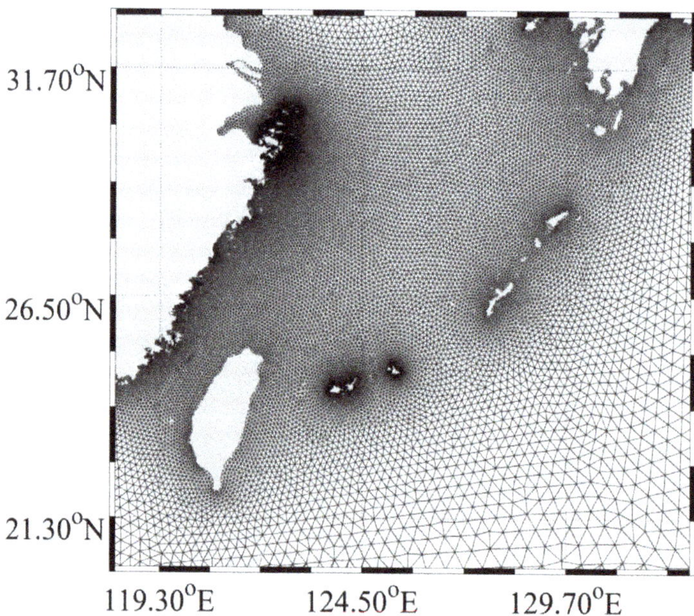

Figure 2. Map of the unstructured grids used in the SWAN and FVCOM simulations.

Table 1. Settings for the Finite-Volume Community Ocean Model (FVCOM).

Forcing field	Holland-European Centre for Medium-Range Weather Forecasts (ECMWF) (H-E) wind with a spatial resolution of 0.125° and time interval of six hours
Open boundary	Water-tide data from TPOX.5; sea-surface temperature, sea-surface salinity, and sea-surface current values at 60-min intervals at a 1/12° (~8 km) grid spatial resolution from the HYbrid Coordinate Ocean Model (HYCOM)
Resolution	Triangular terrain grid with a 60-min temporal resolution

Table 2. Settings for the Simulating Waves Nearshore (SWAN) model.

Forcing field	H-E wind with a spatial resolution of 0.125° and time interval of six hours; current speed simulated by the FVCOM
Frequency bins	Logarithmic, between 0.01 and 1, at intervals of $\Delta f/f = 0.\ 0.903$
Resolution	0.1° grid with a 30-min temporal resolution
Bulk formula	Bulk formula in [39] employed for including the white capping induced by the high wind speed
Directional resolution	10° grid ranged from 0° to 360°
Propagation scheme	WESTHuysen (non-linear saturation-based white-capping combined with wind)
Bottom friction	JONSWAP with a constant friction coefficient
Other settings	Wave interactions (QUADrupl); triad wave–wave interactions (TRIad)

In order to validate the simulated current from the FVCOM, sea-surface-current data from the Climate Forecast System Version 2 (CFSv2) from the National Center of Atmospheric Research (NCAR) with a coarse spatial resolution (~50 km) were used [28,40]. As an example, the map of the CFSv2 current vectors at 06:00 UTC on 11 July 2015 during Typhoon Chan-hom is shown in Figure 3. Although the Kuroshio current in the Taiwan Strait is not very apparent in this figure due to the coarse spatial resolution, the strong wind-driven current with a spiral structure can be seen around the Zhoushan Islands. The FVCOM-simulated sea-water levels during the two typhoons were compared with results from the HYCOM data, which were officially released on the HYCOM's homepage. Figure 4 shows the sea-water-level map from the HYCOM data at 12:00 UTC on 22 September 2014 during Typhoon Fung-wong, exhibiting the storm-surge pattern around the Zhoushan Islands. Moreover, a cyclonic pattern exists, as denoted by the black ellipse in Figure 4, which could be an indicator of offshore upwelling off eastern Guangdong Province, China [41]. In addition, wave measurements from the Jason-2 altimeter mission, which has proven to be good for wave monitoring over global seas [42], were collected during the simulation periods in order to validate the model-simulated waves, i.e., those generated by the WW3 [43] and SWAN [44] models. Figure 5 shows the footprints from the Jason-2 altimeter overlaid on the water-depth map of September 2014, from which the altimeter significant wave height (SWH) data covering the red square were used to validate the simulations by the SWAN model.

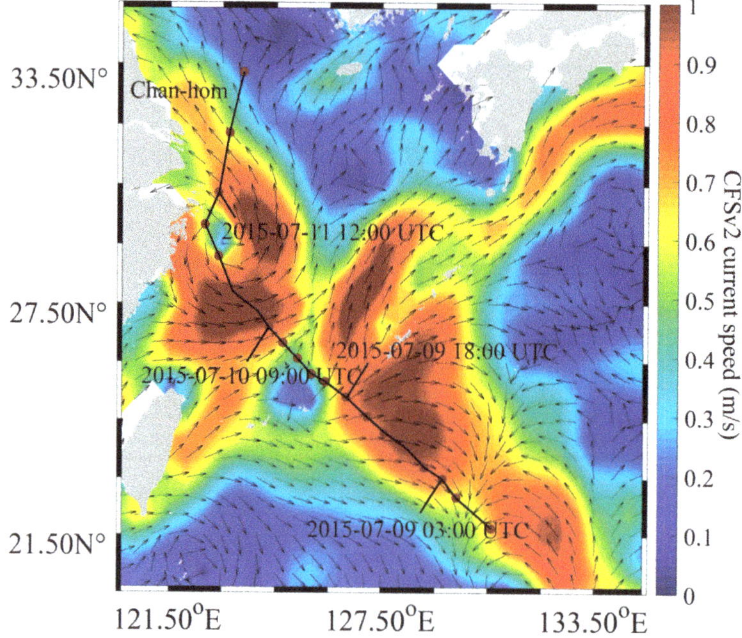

Figure 3. Sea-surface-current vector map from the Climate Forecast System Version 2 (CFSv2) model at 06:00 UTC on 11 July 2015, in which the black line is the track of Typhoon Chan-hom.

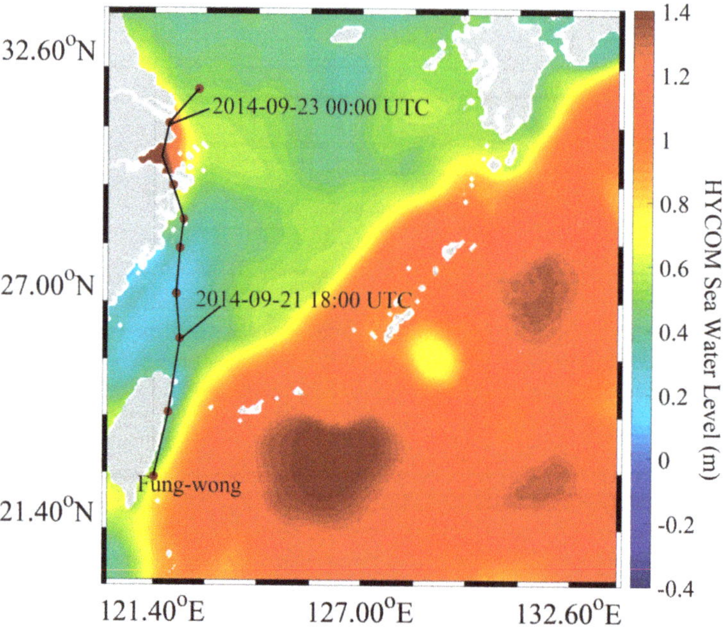

Figure 4. Sea-water-level map from the HYbrid Coordinate Ocean Model (HYCOM) at 12:00 UTC on 22 September 2014, in which the black line is the track of Typhoon Fung-wong.

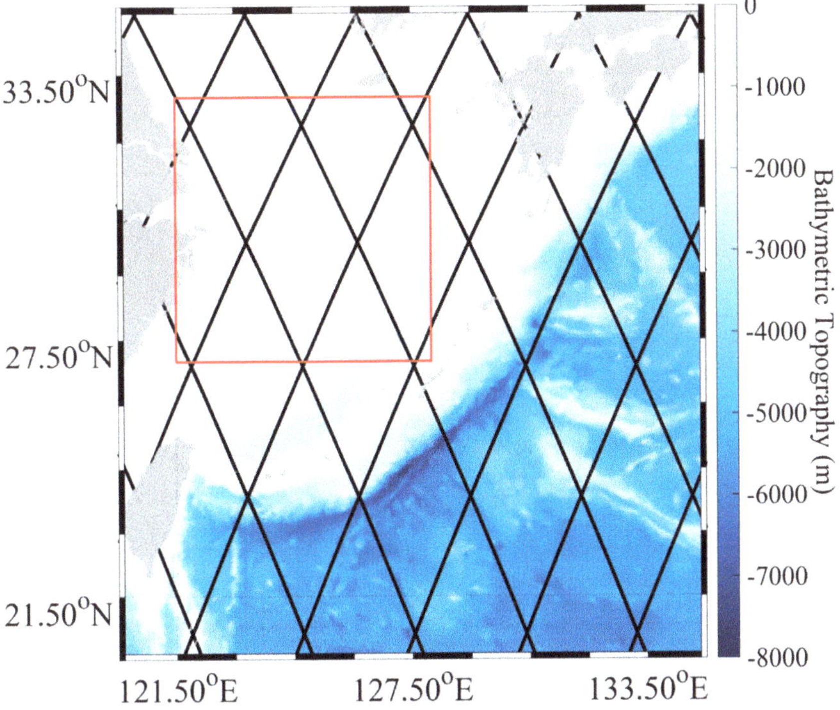

Figure 5. Footprints from the Jason-2 altimeter overlaid on the water-depth map of September 2014, in which the red square encloses the waters around the Zhoushan Islands.

3. Results

In order to confirm the FVCOM output, its simulations of currents and sea-water levels were visualized and then compared with the CFSv2 and HYCOM data, respectively. In addition, waves from the SWAN model for the two typhoons were validated against the satellite altimeter observations.

3.1. Simulation of Sea-Surface Current

Figure 6 shows the current maps from the FVCOM simulation for the periods of Typhoon Fung-wong and Typhoon Chan-hom, respectively. These two cases are located in the region around the Zhoushan Islands, corresponding to the red square in Figure 5. Figure 6a was captured at 13:00 UTC on 22 September 2014, and Figure 6b, at 06:00 UTC on 11 July 2015, when the typhoons were close to the Zhoushan Islands [26]. It is observed that the current speed away from the land increased significantly due to strong winds, reaching 1 m/s. In the region far from the Yangtze Estuary in Figure 6a, it can be seen that the current exhibited an approximate spiral pattern, which was induced by the cyclonic winds associated with the typhoon path in Figure 1a. This type of pattern is also clearly observed in Figure 6b, which is south of the Zhoushan Islands. Moreover, the currents (presumably due to the tides [45]) are well simulated around the Zhoushan Islands due to the fine spatial resolution of the triangular grid, although these currents are not apparent in Figure 3. To confirm the accuracy of the currents in the FVCOM model, CFSv2 data were directly employed. We selected three typical regions around the Zhoushan Islands to statistically analyze the current speed, as presented in Figure 7. Note that Region A is located in the Yangtze Estuary and tidal flats where the tidal current is relatively strong. In particular, three representative specific locations were also selected for studying the extreme waves later. Figure 8 shows the comparisons at Regions A to C, revealing that the RMSE of the current speed was around 0.6 m/s, with a correlation (Cor) more than 0.8. It was also found that the simulated current speeds from the FVCOM were generally higher than those from the CFSv2, with values > 0.3 m, probably caused by the typhoons. We believe that this sort of behavior is reasonable since the composite H-E winds were taken as the forcing field, thus indicating that the underestimation of the ECMWF winds was improved [46]. In particular, the maximum current speeds occur in the Yangtze Estuary and tidal flats, indicating that the tidal current is relatively strong in such regions [47].

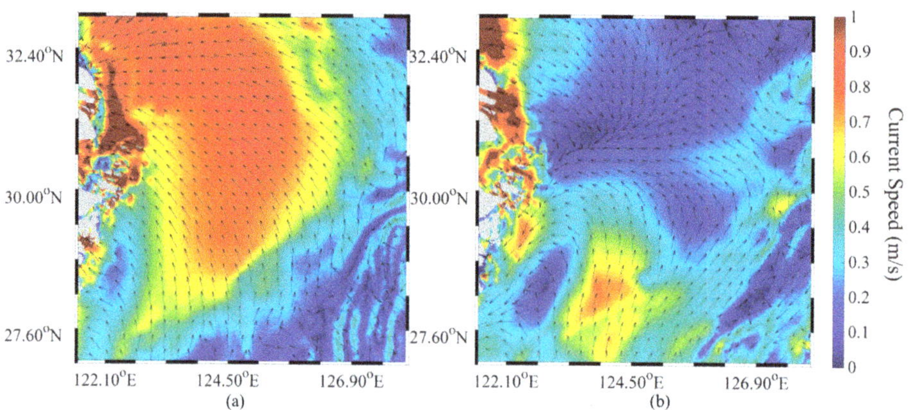

Figure 6. Current-speed maps from the FVCOM simulation at (**a**) 13:00 UTC on 22 September 2014 and (**b**) 06:00 UTC on 11 July 2015.

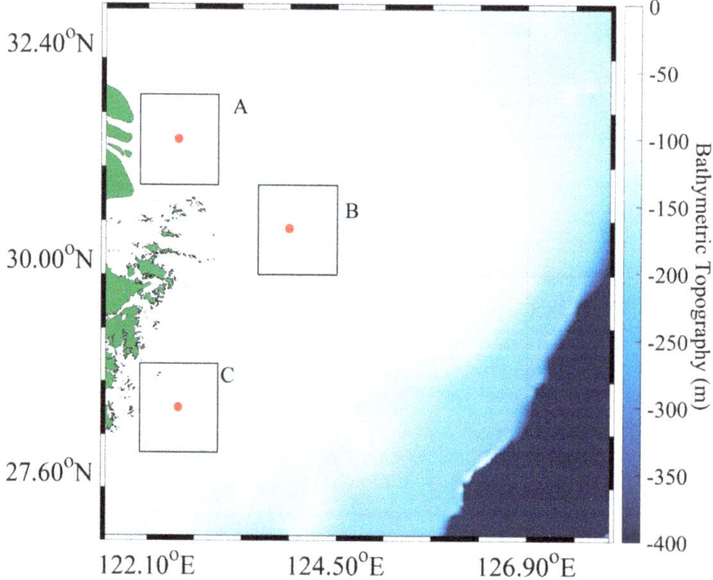

Figure 7. Three typical regions around the Zhoushan Islands used to study the statistical analysis of current speed. Region A is located in the Yangtze Estuary and tidal flats. The three red spots represent the three representative specific locations for studying the extreme waves.

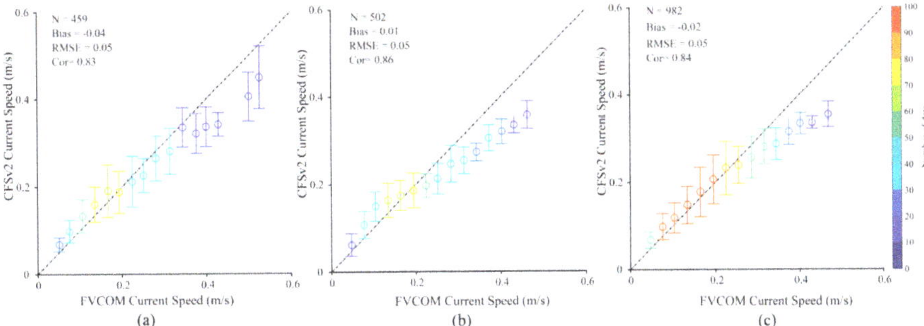

Figure 8. Comparison between current speeds simulated by the FVCOM and the CFSv2 data for a 0.1-m/s bin between 0 and 0.6 m, in which the error bars represent the standard deviations of each bin for the matchups in (**a**) Region A, (**b**) Region B, and (**c**) Region C.

3.2. Simulation of Sea-Water Level

Figure 9 presents the sea-water level maps simulated by the FVCOM at 13:00 UTC on 22 September 2014 (Figure 9a) and 06:00 UTC on 11 July 2014 (Figure 8b). It can be observed that strong winds raise the sea-water level around the Zhoushan Islands, indicating a rise of more than 1 m for the two cases. Interestingly, the sea-water level decreased slightly (~0.2 m) and exhibited a spiral structure, as indicated by the red ellipse in Figure 9b. This pattern is similar to the cyclonic decrease in sea-water level around the Pearl River Estuary shown in Figure 4, and could be the result of either Ekman pumping induced by typhoons or a local mesoscale eddy. This phenomenon is worthy of further study in the future. Unfortunately, independent data, including sea-water levels, are unavailable for public investigators. Therefore, HYCOM sea-water level data were compared with the FVCOM simulations. Figure 10 shows the statistical comparison of the sea-water levels at Regions A to C in Figure 7, revealing an RMSE of around 0.13 m, with a Cor of about 0.86. Thus, the sea-water level simulated by the FVCOM was found to be suitable for conducting the sensitive typhoon-wave-simulation experiment.

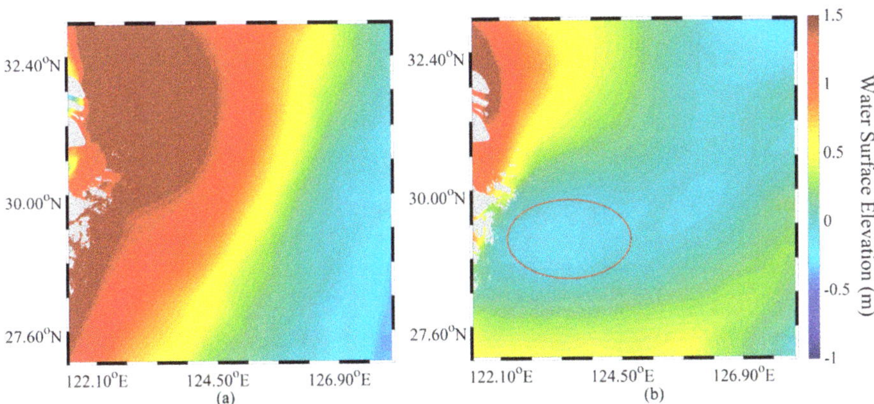

Figure 9. Sea-water-level maps from the FVCOM at (**a**) 13:00 UTC on 22 September 2014 and (**b**) 06:00 UTC on 11 July 2015.

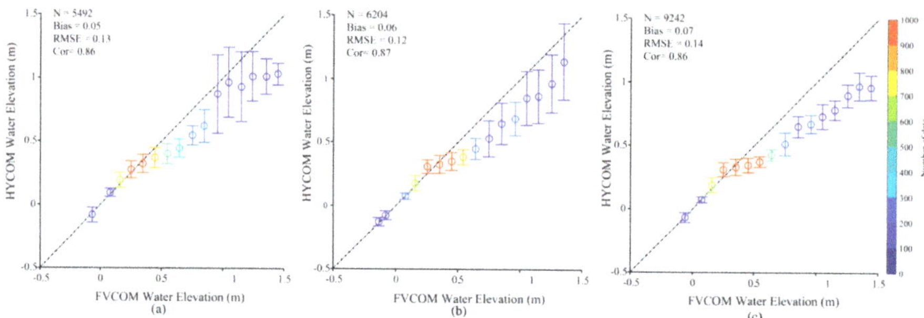

Figure 10. Comparison between sea-water level simulated by the FVCOM and the HYCOM data for a 0.1-m bin between −0.5 and 1.5 m, in which the error bar represents the standard deviation at each bin for the matchups in (**a**) Region A; (**b**) Region B; and (**c**) Region C.

3.3. Simulation of Significant Wave Height

Figure 11 shows the simulated wave maps from the SWAN model considering one condition. It can be seen that the SWH reached 10 m at 14:00 UTC on 22 September 2014 during Typhoon Fung-wong, in which the cyclone center was located over the main island of the Zhoushan Islands. The cyclonic pattern was apparent on the high seas in the East China Sea, although the difference in the coastal regions was significant, e.g., in the Yangtze Estuary and the tidal flats around the Zhoushan Islands. We consistently focused on analyzing the difference between the simulated results from the SWAN model for two conditions, i.e., SWH without the sea-water level minus SWH with the sea-water level. Figure 12 shows that high sea-water level seemed to affect the simulated SWH from the SWAN model; in particular, the difference was up to −0.5 m in the coastal regions due to higher water levels and reduction in breaking/friction. Therefore, we believe that the tidal current and sea-water level play more important roles than the typhoon-induced current in the typhoon-induced wave simulation.

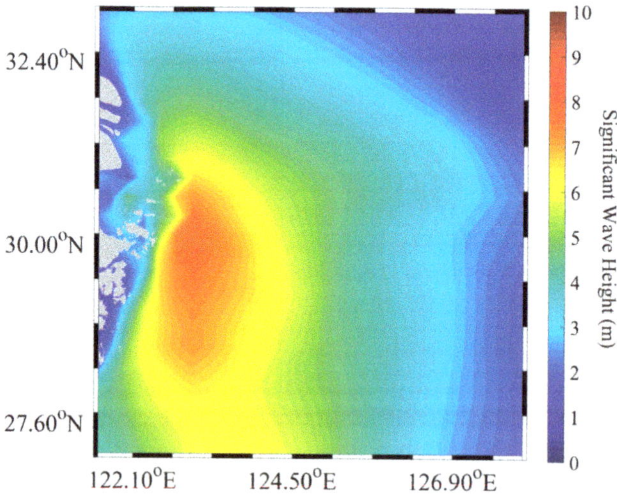

Figure 11. Significant-wave-height maps from the SWAN model at 14:00 UTC on 22 September 2014.

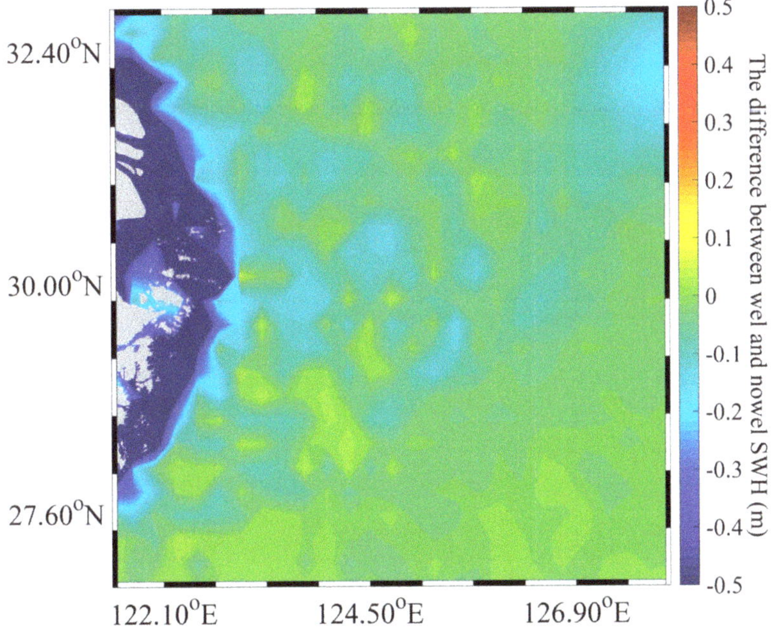

Figure 12. Difference between simulated results from the SWAN model under two conditions (significant wave height (SWH) without sea-water level minus SWH with sea-water level) at 14:00 UTC on 22 September 2014.

Measurements were collected from the altimeter Jason-2 with spatial coverage around the Zhoushan Islands, as marked by the red rectangle in Figure 5. The time series of the SWH from the SWAN model and buoy ID:B14 during typhoon Fung-wong is shown in Figure 13, and it was found that the SWAN simulations were consist with the measurements from the buoy. We compared the simulated results from the SWAN model with the measurements for more than 15,000 matchups. The statistical analysis is plotted in Figure 14. The RMSE was 0.95 m, and the Cor was 0.84 (Figure 14b)—lower than the 1.21 m RMSE and 0.75 Cor when the sea-water level was not included (Figure 14a). Moreover, overestimation was clearly apparent in the high-sea state (SWH > 6 m) without the sea-water level term, while considering the latter significantly improved the wave simulation's performance. Moreover, there seems to be an underprediction by the SWAN model in the low-sea state (SWH < 2 m), which is probably caused by low winds in the H-E model. Meanwhile, the simulated waves were also validated against the measurements from the five moored buoys during the periods 21–23 September 2014 and 10–13 July 2015, as listed in Table 3. Note that measurements from ID: B11 were not available during Typhoon Chan-hom. In general, the results were also consistent with the buoy measurements, indicating SWHs with a ~0.8-m RMSE and Cor > 0.8. Collectively, these findings suggest that the sea-water level should be included in the wave simulation using the SWAN model for typhoons and hurricanes, since it modulates waves in both low- and high-sea states.

Table 3. Comparison between the results from the SWAN model and the moored buoys.

	ID: B11	ID: B12	ID: B13	ID: B14	ID: B15	ID: B20
RMSE	0.84	0.64	0.79	0.84	0.88	0.75
Cor	0.92	0.93	0.84	0.94	0.85	0.94
Bias	0.44	0.19	0.75	0.19	−0.21	0.31

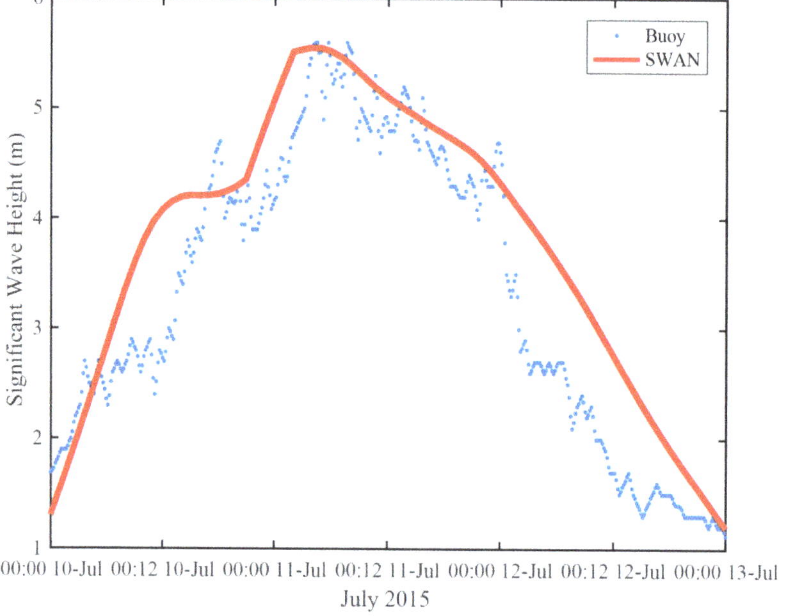

Figure 13. Comparison between significant wave heights simulated by the SWAN model and those recorded from the buoy B14 during Typhoon Chan-hom.

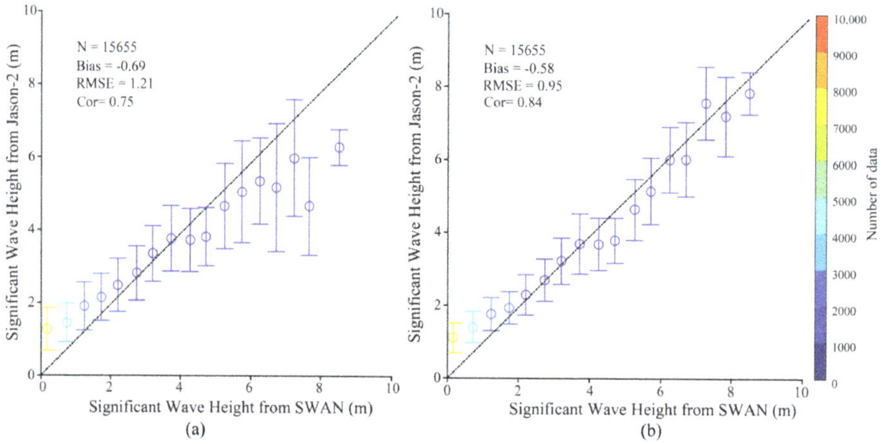

Figure 14. Comparison between significant wave heights simulated by the SWAN model and the measurements from the Jason-2 altimeter for a 0.14-m bin between 0 and 9 m, in which the error bars represent the standard deviation at each bin: (**a**) without the sea-water level term; (**b**) with the sea-water level term.

4. Discussion

We further investigated the characteristics of the typhoon-induced waves when Typhoon Fung-wong passed the Zhoushan Islands. In particular, three points in Regions A to C were selected, which are marked in Figure 7. The average SWH maps simulated by the SWAN model during the period from 20 to 23 September 2014 are shown in Figure 15; the maximum SWH was 6 m, but the high-sea state induced by the typhoon waves seems to have been blocked outside of the Zhoushan Islands when the typhoon track encountered the islands. This kind of behavior is apparent on 22 September 2014. We think that the barrier of islands results in the low-sea state in typhoons.

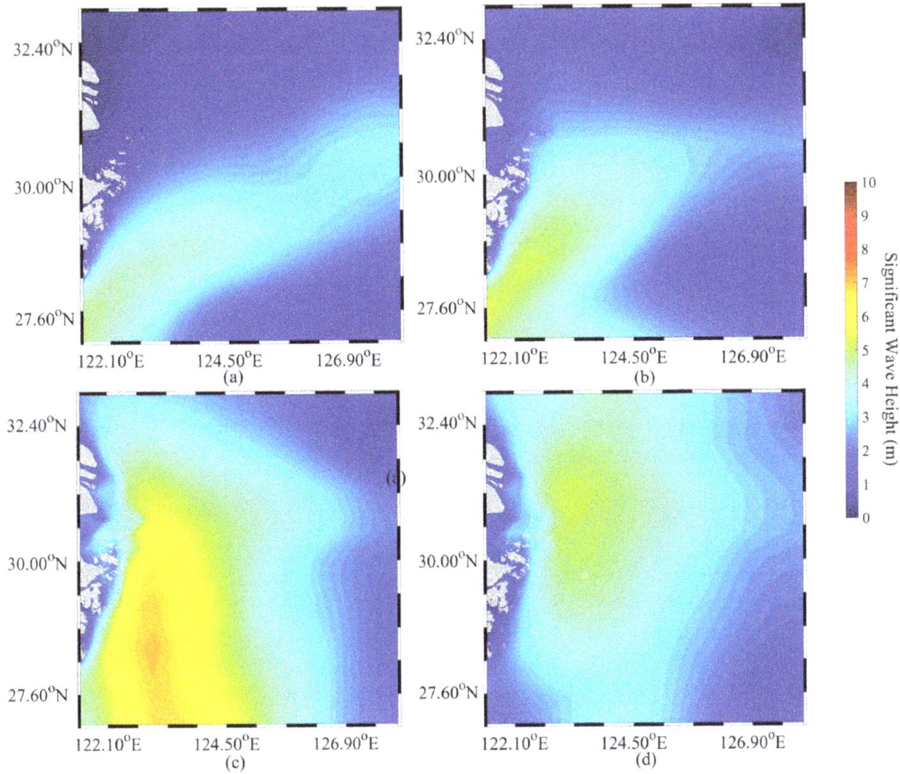

Figure 15. The average SWH maps simulated by the SWAN model when Typhoon Fung-wong passed the Zhoushan Islands on (**a**) 20 September 2014, (**b**) 21 September 2014, (**c**) 23 September 2014, and (**d**) 23 September 2014.

Figure 16 shows the time series of the simulated SWH from the SWAN model during the period from 20 September to 24 September 2014 corresponding to the three points A–C in Figure 7; the maximum SWH was 7 m on 22 September when the typhoon's path crossed the islands. It is not surprising that the SWH at Points A and B is smaller than that at Point C, because Point C is located in open sea and the sea-water levels should modify the waves at Points A and B in the Yangtze Estuary and tidal flats. It is also clearly observed that the tendency of the SWH at Points A and B accords with the time series. Especially during the period from 00:00 20 September to 12:00 20 September, the SWH decreases at Points A and B, which is probably caused by the change in sea-water levels, while the SWH decreases at Point C. Interestingly, low SWH is achieved at Point A, e.g., there was a reduction of more than 1m on 23 September, when the maximum SWH occurred.

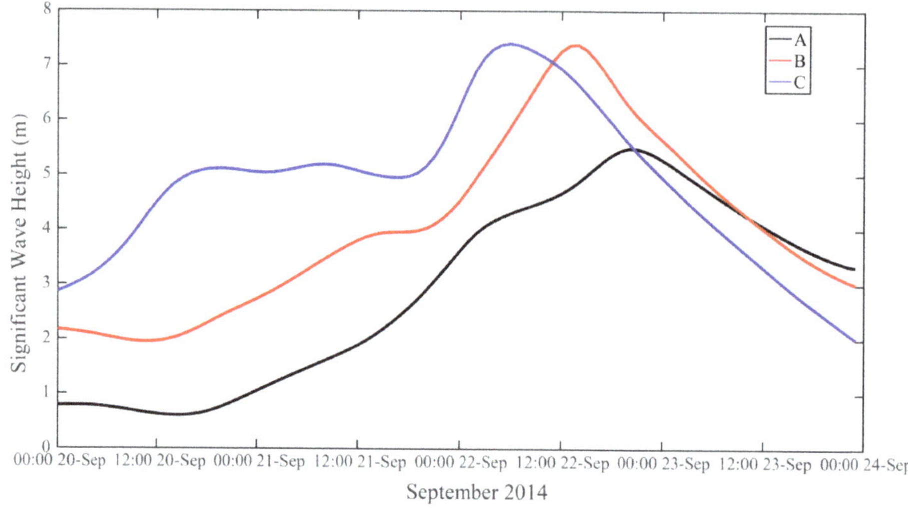

Figure 16. The time series of the simulated SWH at the selected Points A to C shown in Figure 7 during the period from 20 September to 24 September 2014.

5. Conclusions

The tidal currents are often strong, and the sea-water levels vary in some areas, e.g., the Yangtze Estuary, Hangzhou Bay, and the Zhoushan Islands. In our previous study [26], typhoon waves were simulated by the WAVEWATCH-III model around the Zhoushan Islands, although these were simulated without considering the influence of currents and sea-water levels during Typhoon Fung-wong (2014) and Typhoon Chan-hom (2015). The effect of typhoon-induced currents and the Kuroshio Current on typhoon-induced wave simulation has been frequently studied, and it has been concluded that SWH is significantly correlated with wind-induced-current speed when the current speed is >0.5 m/s [28]. The main purpose of this study was to investigate the influence of sea-water levels on wave simulation using the SWAN model and the FVCOM in the shallow waters around the Zhoushan Islands. These two numerical ocean models share a unique unstructured grid of the same simulated region. The composited H-E wind was utilized as the forcing field when running the model. The validation of H-E winds against the measurements from moored buoys has been discussed in [26] and could ameliorate the underestimation of ECMWF winds. Specifically, TPXO.5 tide data and HYCOM sea-surface-temperature, sea-surface salinity, and sea-surface-current data were also collected, which were taken as the open-boundary conditions. The comparison of the simulated current speeds from the FVCOM with those from the CFSv2 revealed a 0.12-m RMSE with a 0.73 Cor. Similarly, the HYCOM sea-water level data were used to confirm the accuracy of the simulations from the FVCOM, yielding a 0.2-m RMSE and 0.73 Cor.

Waves were simulated by the SWAN model under two conditions: (1) including the wind and current, and (2) including the wind, current, and sea-water level. The pattern of differences between them (up to −0.5 m) was consistent with the pattern of the simulated wave height from the FVCOM in coastal regions, e.g., in the Yangtze Estuary and tidal flats around the Zhoushan Islands. The validation of the SWH with the sea-water level against Jason-2 altimeter data and moored buoy measurements demonstrated an RMSE < 0.9 m with a Cor > 0.8. In particular, the overestimation in the high-sea state (SWH > 6 m) without the sea-water level term was ameliorated. Therefore, we conclude that the sea-water level should be in included in typhoon-induced wave simulations with the SWAN model.

Three representative specific locations were selected to study the characteristics of extreme waves around the Zhoushan Islands during Typhoon Fung-wong (2014). It was found that the SWHs at

Points A and B located in the Yangtze Estuary and tidal flats were generally smaller than the SWH at Point C in open sea due to the barrier of islands. In particular, the reduction of SWH was up to 1 m on 23 September 2014, when the maximum waves occurred. As a matter of fact, mesoscale eddies and upwelling are common in coastal waters. In the near future, we plan to study the coastal oceanic feedbacks of typhoons, e.g., Ekman pumping and sea-surface-temperature cooling.

Author Contributions: Conceptualization, W.S. and Z.Y.; methodology, W.S., Z.Y., and Y.D.; validation, Z.Y., Y.D., and J.S.; formal analysis, W.S. and J.S.; investigation, Z.Y.; resources, Y.D.; writing—original draft preparation, W.S.; writing—review and editing, J.S.; visualization, Z.Y. and Q.J.; funding acquisition, W.S. and Q.J. All authors have read and agreed to the published version of the manuscript.

Funding: This research was funded by the National Key Research and Development Program of China under contract nos. 2017YFA0604901 and 2016YFC1401605, the National Natural Science Foundation of China under contract nos. 41806005 and 41776183, and the Science and Technology Project of Zhoushan City, China, under contract no. 2019C21008.

Acknowledgments: We appreciate the Simulating WAves Nearshore (SWAN) model developed at the Delft University of Technology. We also thank the Marine Ecosystem Dynamics Modeling Laboratory (MEDML) for providing the Finite-Volume Community Ocean Model (FVCOM) in detail. The European Centre for Medium-Range Weather Forecasts (ECMWF)'s wind data were accessed via http://www.ecmwf.int. The tide data were downloaded from https://www.tpxo.net. The sea-surface temperature, sea-surface-salinity, sea-surface-current, and sea-water-level data from the HYbrid Coordinate Ocean Model (HYCOM) were obtained via https://www.hycom.org. Sea-surface-current data from the National Centers for Environmental Prediction (NCEP) Climate Forecast System Version 2 (CFSv2) were obtained via http://cfs.ncep.noaa.gov. The General Bathymetric Chart of the Oceans (GEBCO) data were downloaded from ftp.edcftp.cr.usgs.gov. The buoy measurements were kindly shared by Dr. Shuiqing Li at the Institute of Oceanology, Chinese Academy of Sciences.

Conflicts of Interest: The authors declare no conflict of interest.

Appendix A

The governing equation of the SWAN model considering ambient current is determined by the evolution of the wave spectrum and expressed as follows:

$$\frac{\partial N}{\partial t} + \Delta \cdot \left[(C_g + V)N \right] + \frac{\partial C_\omega N}{\partial \omega} + \frac{\partial C_\theta N}{\partial \theta} = \frac{S_t}{\omega}, \tag{A1}$$

in which N is the wave-action-density spectrum; t is the time; ω is the wave frequency; θ is the wave-propagation direction; C_ω and C_θ are the wave-propagation velocities in terms of the spectral space ω and θ, respectively; C_g is the group-velocity vector; V is the sea-surface-current vector; and Δ is the Hamiltonian divergence operator. S_t comprises the input and dissipation source terms, stated as

$$S_t = S_{in} + S_{bot} + S_{nl} + S_{tq} + S_{db}, \tag{A2}$$

where S_t includes an atmosphere–wave interaction term of the function for the wind-induced wave growth S_{in}; S_{bot} is the friction induced by wave–bottom interaction; S_{nl} is the nonlinear wave–wave interaction term; S_{tq} is the three- (triad) and four-wave components (quadruplets) of the wave–wave interactions; and S_{db} is the wave decay due to white capping and depth-induced wave breaking. More detailed descriptions of these terms are provided in the technical manual of the SWAN model.

Appendix B

The governing equations of the FVCOM consist of the following momentum, continuity, temperature, salinity, and density expressions:

$$\frac{\partial u}{\partial t} + u\frac{\partial u}{\partial x} + v\frac{\partial u}{\partial y} + w\frac{\partial u}{\partial z} - fv = -\frac{1}{\rho} + \frac{\partial(P_H + P_a)}{\partial x} - \frac{1}{\rho}\frac{\partial q}{\partial x} + \frac{\partial}{\partial z}\left(K_m\frac{\partial u}{\partial z}\right) + F_u, \tag{A3}$$

$$\frac{\partial v}{\partial t} + u\frac{\partial v}{\partial x} + v\frac{\partial v}{\partial y} + w\frac{\partial v}{\partial z} - fu = -\frac{1}{\rho} + \frac{\partial(P_H + P_a)}{\partial y} - \frac{1}{\rho}\frac{\partial q}{\partial y} + \frac{\partial}{\partial z}\left(K_m\frac{\partial v}{\partial z}\right) + F_v, \tag{A4}$$

$$\frac{\partial w}{\partial t} + u\frac{\partial w}{\partial x} + v\frac{\partial w}{\partial y} + w\frac{\partial w}{\partial z} = -\frac{1}{\rho}\frac{\partial q}{\partial y} + \frac{\partial}{\partial z}\left(K_m \frac{\partial w}{\partial z}\right) + F_w, \tag{A5}$$

$$\frac{\partial u}{\partial x} + \frac{\partial v}{\partial y} + \frac{\partial w}{\partial z} = 0, \tag{A6}$$

$$\frac{\partial T}{\partial t} + u\frac{\partial T}{\partial x} + v\frac{\partial T}{\partial y} + \frac{\partial T}{\partial z} = \frac{\partial}{\partial z}\left(K_h \frac{\partial T}{\partial z}\right) + F_T, \tag{A7}$$

$$\frac{\partial S}{\partial t} + u\frac{\partial S}{\partial x} + v\frac{\partial S}{\partial y} + \frac{\partial S}{\partial z} = \frac{\partial}{\partial z}\left(K_h \frac{\partial S}{\partial z}\right) + F_S, \tag{A8}$$

$$\rho = \rho(T, S, P), \tag{A9}$$

where x, y, and z are the east, north, and vertical axes in the Cartesian coordinate system, respectively; u, v, and w are the x, y, and z velocity components, respectively; T is the sea temperature; S is the salinity; ρ is the sea-water density; p_a is the air pressure at the sea surface; p_H is the hydrostatic pressure; q is the non-hydrostatic pressure; f is the Coriolis parameter; g is the acceleration of gravity; K_m is the vertical-eddy-viscosity coefficient; and K_h is the thermal-vertical-eddy-diffusion coefficient. F_u, F_v, and F_w represent the horizontal momentum in the x direction, horizontal momentum in the y direction, and vertical momentum in the z direction, respectively; F_T and F_S represent the thermal and salt diffusion terms, respectively. The total water-column depth is $D = H + \zeta$, in which H is the bottom depth (relative to $z = 0$) and ζ is the height of the free surface (relative to $z = 0$). $p = p_a + p_H + q$ is the total pressure, in which the hydrostatic pressure p_H satisfies

$$\frac{\partial p_H}{\partial z} = -\rho g \Longrightarrow p_H = \rho_o g\zeta + g\int_z^0 \rho dz'. \tag{A10}$$

These terms are described in further detail in the user manual of the FVCOM.

References

1. Buonaiuto, F.S.; Slattery, M.; Bokuniewicz, H.J. Wave modeling of long island coastal waters. *J. Coastal Res.* **2011**, *27*, 470–477.
2. Wang, Z.F.; Zhou, L.B.; Dong, S.; Wu, L.Y.; Li, Z.B.; Mu, L.; Wang, A.F. Wind wave characteristics and engineering environment of the South China Sea. *J. Ocean U. China* **2014**, *13*, 893–900. [CrossRef]
3. Liu, Q.X.; Babanin, A.; Fan, Y.; Zieger, S.; Guan, C.L.; Moon, I.J. Numerical simulations of ocean surface waves under hurricane conditions: Assessment of existing model performance. *Ocean Model.* **2017**, *18*, 73–79. [CrossRef]
4. Wang, N.; Hou, Y.J.; Li, S.Q.; Li, R. Numerical simulation and preliminary analysis of typhoon waves during three typhoons in the Yellow Sea and east China Sea. *J. Oceanol. Limn.* **2019**, *37*, 1805–1816. [CrossRef]
5. Zhou, L.M.; Li, Z.B.; Mu, L.; Wang, A.F. Numerical simulation of wave field in the South China Sea using WAVEWATCH III. *Chin. J. Oceanol. Limn.* **2014**, *37*, 656–664. [CrossRef]
6. Rogers, W.E.; Hwang, P.A.; Wang, D.W. Investigation of wave growth and decay in the SWAN model: Three regional-scale applications. *J. Phys. Oceanogr.* **2003**, *33*, 366–389. [CrossRef]
7. Bi, F.; Song, J.B.; Wu, K.J.; Xu, Y. Evaluation of the simulation capability of the Wavewatch III model for Pacific Ocean wave. *Acta Oceanol. Sin.* **2015**, *34*, 43–57. [CrossRef]
8. Mentaschi, L.; Besio, G.; Cassola, F.; Mazzino, A. Performance evaluation of Wavewatch IIIi in the Mediterranean Sea. *Ocean Modell.* **2015**, *90*, 82–94. [CrossRef]
9. Akpınar, A.; Vledder, G.; Kömürcü, M.; Özger, M. Evaluation of the numerical wave model (SWAN) for wave simulation in the Black Sea. *Cont. Shelf Res.* **2012**, *50*, 80–99.
10. Rusu, L.; Soares, C.G. Wave energy assessments in the Azores islands. *Renew. Energ.* **2012**, *45*, 183–196. [CrossRef]
11. Bottema, M.; Vledder, G.V. Effective fetch and non-linear four-wave interactions during wave growth in slanting fetch conditions. *Coast. Eng.* **2008**, *55*, 261–275. [CrossRef]

12. Dragani, W.C.; Garavento, E.; Simionato, C.G.; Nunez, M.N.; Martin, P.; Campos, M.I. Wave simulation in the outer Rio de la Plata estuary: Evaluation of SWAN model. *J. Waterw. Port C-ASCE.* **2008**, *5*, 299–305. [CrossRef]

13. Moeini, M.H.; Etemad-Shahidi, A. Wave parameter hindcasting in a lake using the SWAN model. *Sci. Iran.* **2009**, *16*, 156–164.

14. Young, I.R.; Vledder, G. A review of the central role of nonlinear interactions in wind-wave evolution. *Phil. Trans. Roy. Soc. Lond.* **1993**, *342*, 505–524.

15. Young, I.R. Observations of the spectra of hurricane generated waves. *Ocean Eng.* **1998**, *25*, 261–276. [CrossRef]

16. Cui, H.; He, H.L.; Liu, X.H.; Li, Y. 2012. Effect of oceanic current on typhoon-wave modeling in the East China Sea. *Chin. Phys. B.* **2012**, *21*, 109201. [CrossRef]

17. Sheng, Y.P.; Alymov, V.; Paramygin, V.A. Simulation of storm surge, wave, currents, and inundation in the outer banks and Chesapeake Bay during Hurricane Isabel in 2003: The importance of waves. *J. Geophys. Res.* **2010**, *115*, C04008. [CrossRef]

18. Dietrich, J.C.; Tanaka, S.; Westerink, J.J.; Dawson, C.N.; Luettich, R.A.; Zijlema, M.; Holthuijsen, L.H.; Smith, J.M.; Westerink, L.G.; Westerink, H.J. Performance of the unstructured-mesh, SWAN+ ADCIRC model in computing hurricane waves and surge. *J. Sci. Comput.* **2012**, *52*, 468–497. [CrossRef]

19. Chen, C.S.; Liu, H.; Beardsley, R.C. An unstructured, finite-volume, three-dimensional, primitive equation ocean model: Application to coastal ocean and estuaries. *J. Atmos. Ocean. Tech.* **2003**, *20*, 159–186. [CrossRef]

20. Pringle, J.M. Sources of variability in Gulf of Maine circulation, and the observations needed to model it. *Deep-Sea Res. PT. II.* **2006**, *53*, 2457–2476. [CrossRef]

21. Huang, H.; Chen, C.L.; Cowles, G.W.; Winant, C.D.; Beardsley, R.C.; Hedstrom, K.S. Haidvogel, D.B. FVCOM validation experiments: Comparisons with ROMS for three idealized barotropic test problems. *J. Geophys. Res.* **2008**, *113*, C07042. [CrossRef]

22. Chen, C.S.; Huang, H.; Beardsley, R.C.; Liu, H.; Xu, Q.; Cowles, G. A finite-volume numerical approach for coastal ocean circulation studies: Comparisons with finite difference models. *J. Geophys. Res.* **2007**, *112*, C03018. [CrossRef]

23. Ma, Z.; Han, G.; De Young, B. Oceanic responses to Hurricane Igor over the Grand Banks: A modeling study. *J. Geophys. Res.* **2015**, *120*, 1276–1295. [CrossRef]

24. Sun, Y.J.; Perrie, W.; Toulany, B. Simulation of wave-current interactions under hurricane conditions using an unstructured-grid model: Impacts on ocean waves. *J. Geophys. Res.* **2018**, *123*, 3739–3760. [CrossRef]

25. Dukhovskoy, D.S.; Morey, S.L. Simulation of the Hurricane Dennis storm surge and considerations for vertical resolution. *Nat. Hazards* **2011**, *58*, 511–540. [CrossRef]

26. Sheng, Y.X.; Shao, W.Z.; Li, S.Q.; Zhang, Y.M.; Yang, H.W.; Zuo, J.C. Evaluation of typhoon waves simulated by WaveWatch-III model in shallow waters around Zhoushan Islands. *J. Ocean U. China* **2019**, *18*, 365–375. [CrossRef]

27. Shao, W.Z.; Sheng, Y.X.; Li, H.; Shi, J.; Ji, Q.Y.; Tan, W.; Zuo, J.C. Analysis of wave distribution simulated by WAVEWATCH-III model in typhoons passing Beibu Gulf, China. *Atmosphere* **2018**, *9*, 265. [CrossRef]

28. Hu, Y.Y.; Shao, W.Z.; Shi, J.; Sun, J.; Ji, Q.Y.; Cai, L.N. Analysis of the typhoon wave distribution simulated in WAVEWATCH-III model in the context of Kuroshio and wind-induced current. *J. Oceanol. Limn.* **2020**. [CrossRef]

29. Moon, I.J. Impact of a coupled ocean wave-tide-circulation system on coastal modeling. *Ocean Model.* **2005**, *8*, 203–236. [CrossRef]

30. Masson, D. A case study of wave-current interaction in a strong tidal current. *J. Phys. Oceanogr.* **1996**, *26*, 359–372. [CrossRef]

31. Chen, T.Q.; Zhang, Q.H.; Wu, Y.S.; Ji, C.; Yang, J.S.; Liu, G.W. Development of a wave-current model through coupling of FVCOM and SWAN. *Ocean Eng.* **2018**, *164*, 443–454. [CrossRef]

32. Lv, X.; Yuan, D.; Ma, X.; Tao, J. Wave characteristics analysis in Bohai Sea based on ECMWF wind field. *Ocean Eng.* **2014**, *91*, 159–171. [CrossRef]

33. Kamranzad, B.; Etemad-Shahidi, A.; Chegini, V.; Yeganeh-Bakhtiary, A. Climate change impact on wave energy in the Persian Gulf. *Ocean Dynam.* **2015**, *65*, 777–794. [CrossRef]

34. Zheng, K.W.; Sun, J.; Guan, C.L.; Shao, W.Z. Analysis of the global swell and wind-sea energy distribution using WAVEWATCH III. *Adv. Meteorol.* **2016**, *7*, 1–9. [CrossRef]

35. Zheng, K.W.; Osinowo, A.; Sun, J.; Hu, W. Long term characterization of sea conditions in the East China Sea using significant wave height and wind speed. *J. Ocean U. China* **2018**, *17*, 733–743. [CrossRef]
36. Hu, Y.Y.; Shao, W.Z.; Wei, Y.L.; Zuo, J.C. Analysis of typhoon-induced waves along typhoon tracks in the Western North Pacific Ocean, 1998–2017. *J. Mar. Sci. Eng.* **2020**, *8*, 521. [CrossRef]
37. Wang, K.; Hou, Y.J.; Li, S.Q.; Li, R. Numerical study of storm surge inundation in the southwestern Hangzhou bay region during Typhoon Chan-Hom in 2015. *J. Ocean U. China* **2020**, *19*, 263–271. [CrossRef]
38. Ding, Y.; Yao, Z.G.; Zhou, L.L.; Bao, M.; Zang, Z.C. Numerical modeling of the seasonal circulation in the coastal ocean of the Northern South China Sea. *Front. Earth Sci.* **2018**, *14*, 90–109. [CrossRef]
39. Ou, S.H.; Liau, J.M.; Hsu, T.W.; Tzang, S.Y. Simulating typhoon waves by SWAN wave model in coastal waters of Taiwan. *Ocean Eng.* **2002**, *29*, 947–971. [CrossRef]
40. Yuan, X.; Wood, E.F.; Luo, L.F.; Pan, M. A first look at Climate Forecast System Version 2 (CFSv2) for hydrological seasonal prediction. *Geophys. Res.Lett.* **2011**, *38*, L13402. [CrossRef]
41. Zhuang, W.; Wang, D.X.; Wu, R.S. Coastal upwelling off eastern Fujian\\Guangdong detected by remote sensing. *Chin. J. Atmos. Sci.* **2005**, *29*, 438–444.
42. Zhang, H.F.; Wu, Q.; Chen, G. Validation of HY-2A remotely sensed wave heights against buoy data and Jason-2 altimeter measurements. *J. Atmos. Ocean. Tech.* **2015**, *32*, 1270–1280. [CrossRef]
43. Chu, P.C.; Qi, Y.; Chen, Y.; Shi, P.; Mao, Q. South China Sea wind-wave characteristics. Part I: Validation of Wavewatch-III using TOPEX/Poseidon data. *J. Atmos. Ocean. Tech.* **2004**, *21*, 1718–1733. [CrossRef]
44. Lin, Y.; Dong, S.; Wang, Z.; Guedes Soares, C. Wave energy assessment in the china adjacent seas on the basis of a 20-year SWAN simulation with unstructured grids. *Renew. Energ.* **2019**, *136*, 275–295. [CrossRef]
45. Wu, H.; Du, M.; Wang, X.Y.; Meng, J. Study on hydrography and small-scale process over Zhoushan sea area. *J. Ocean U. China* **2015**, *14*, 829–834. [CrossRef]
46. Stopa, J.E.; Cheung, K.F. Intercomparison of wind and wave data from the ECMWF reanalysis interim and the NECP climate forecast system reanalysis. *Ocean Model.* **2014**, *75*, 65–83. [CrossRef]
47. Shen, Z.B.; Wu, X.B.; Fei, Y.J.; Xu, X.A.; Chen, X.F. Surface tidal currents in the open sea area to the east of the Zhoushan Islands measured with high frequency surface wave radar. *Acta Oceanol. Sin.* **2013**, *32*, 5–10. [CrossRef]

 Journal of
Marine Science and Engineering

Article

Analysis of Typhoon-Induced Waves along Typhoon Tracks in the Western North Pacific Ocean, 1998–2017

Yuyi Hu [1], Weizeng Shao [1,2,*], Yongliang Wei [3] and Juncheng Zuo [1]

[1] Marine Science and Technology College, Zhejiang Ocean University, Zhoushan 316000, China; huyuyiyk@outlook.com (Y.H.); zjuncheng@zjou.edu.cn (J.Z.)
[2] Key Laboratory of Space Ocean Remote Sensing and Application, Ministry of Natural Resources, Beijing 100081, China
[3] College of Marine Sciences, Shanghai Ocean University, Shanghai 201306, China; yl-wei@shou.edu.cn
* Correspondence: shaoweizeng@mail.tsinghua.edu.cn; Tel.: +86-0580-2550-753

Received: 30 June 2020; Accepted: 15 July 2020; Published: 16 July 2020

Abstract: In this study, Version 5.16 of the WAVEWATCH-III (WW3) model is used to simulate parameters of typhoon-generated wave fields in the Western North Pacific Ocean during the period 1998–2017. From a database of more than 300 typhoons, typhoon tracks are partitioned into six groups by their direction of motion and longitude of recurvature track. For typhoons that recurve east of 140° E, or track toward mainland Asia, regions of high significant wave height (SWH) values are separated by a minimum in SWH near 30° N. Partitioning SWH into wind sea and swell components demonstrates that variations in typhoon tracks produce a much stronger signal in the wind sea component of the wave system. Empirical orthogonal function (EOF) analysis is used to compute the four leading modes of variation in average SWH simulated by the WW3 model. The first EOF mode contributes to 17.3% of the total variance; all other modes contribute less than 10%. The first EOF mode also oscillates on an approximately 1-year cycle during the period 1998–2017. Overall, typhoon-induced wave energy dominates north of 30° N. Temporal analysis of the leading principal component of SWH indicates that (a) the intensity of the wave pattern produced by westward-tracking typhoons decreased during the last 20 years, and (b) typhoons that recurve east of 140° E and those that track westward toward southeast Asia are largely responsible for the decadal variability of typhoon-induced wave distribution.

Keywords: wave distribution; typhoon tracks; WAVEWATCH-III; typhoon wave climate; empirical orthogonal function

---

1. Introduction

Tropical cyclones (TCs) play an important role in water vapor and heat transport at the atmosphere–ocean boundary layer, and typhoon landfall is often accompanied by extreme weather such as damaging winds, torrential rain, storm surge, and breaking waves. These typhoon hazards often cause secondary disasters such as floods, landslides, and mudslides. Among ocean basins, the Western North Pacific (WNP) experiences the highest frequency of TC occurrences worldwide. During the last decade, some studies have evaluated the meteorological factors that influence typhoon tracks in this basin (e.g., [1]) and others have used cluster analysis to study typhoon tracks (e.g., [2]). However, the relationship between meteorological factors that influence typhoon tracks and the distribution of typhoon-induced waves remains understudied, especially from a climate perspective. Therefore, the relationship between typhoon track and the spatial and temporal distribution of typhoon-induced waves merits further exploration.

Traditionally, there are two primary methods to analyze the distribution of the typhoon-induced waves: numerical simulations and satellite remote sensing. Frequently used numerical wave simulations include the wave model (WAM), simulating waves nearshore (SWAN) [3], and WAVEWATCH-III (WW3) [4]. The WW3 model has been widely used for wave simulation in many oceanic regions, including the Pacific Ocean [5,6], Indian Ocean [7], China Seas [8,9], and North Atlantic Ocean [10,11]. The WW3 model has been demonstrated to be suitable for studying typhoon-induced waves through validation against moored buoys and satellite altimeter measurements [12–14]. Moreover, the WW3 model is capable of simulating long-period waves for the wave climate analysis [11,15]. Satellite data used for real-time wave observations include altimeter data [16] and synthetic aperture radar (SAR) data [17]. Spaceborne SARs, such as the Chinese Gaofen-3 SAR, feature a large swath width (>500 km) and fine spatial resolution (e.g., 150 m), thereby enabling high-resolution monitoring of typhoon winds [17–19] and waves [20–23] on large spatial scales. However, SAR data are only acquired on demand and are therefore unavailable for long-term studies.

Numerical wave models such as WAM, SWAN, and WW3 are widely used for long-term wave climate analysis under various oceanic and atmospheric conditions [24–27]. These models have been used to analyze the wave climate of the open ocean, coastal regions, and inland seas, including studies in the North Atlantic [27], the Bohai Sea [26], the Yellow Sea [25], the East China Sea [24], and the Black Sea [28]. These numerical wave models have also been used to study the climate of extreme waves [29], including in specific regions such as the China Seas [30], the East China Sea [31], and the Mediterranean Sea [32].

Aspects of numerical wave simulation work focus on evaluating wave climatology in future climates affected by sea-level rise [33] and changes in sea-surface temperatures [34]. Some of these studies have included TC events; for example, Niroomandi et al. (2018) used SWAN to evaluate the wave climate of the Chesapeake Bay, including four TCs or ex-TCs that affected the Bay between 2008 and 2012 [35]. That study, however, analyzed a limited region and a small number of TCs. Therefore, a targeted study of typhoon-induced waves in the WNP is necessary to provide information about the long-term extreme wave climate and the typhoon-induced wave fields produced by various typhoon tracks over the WNP.

In this study, we use the WW3 model to simulate the 1998–2017 Western North Pacific typhoon-induced wave climate. We analyze the wave climate patterns produced by varying typhoon track groups during this period. The remainder of this paper is organized as follows: the setup of the WW3 model is described in Section 2; typhoon track classification and wave analysis corresponding to each track type are described in Section 3. Besides, EOF analysis of significant wave height patterns is also presented; the natural variability of twenty-years' variability in typhoon-induced wave distribution is exhibited in Section 4; and the conclusions are summarized in Section 5.

2. Materials and Methods

We used the WW3 model to simulate typhoon-induced wave parameters (significant wave height (SWH), mean wave period (MWP), and mean wave length (MWL)) in the WNP during the 20-year period 1998–2017. The simulated area is a rectangle from 5° S to 65° N and 93° E to 167° W; bathymetry is provided by the general bathymetry chart of the oceans (GEBCO) at 0.1° horizontal resolution (Figure 1). Data from six National Data Buoy Center (NDBC) buoys in the Western North Pacific Ocean are used to validate the WW3-simulated SWHs. Locations of the NDBC buoys are marked Figure 1. Typhoon track data are provided by the Regional Specialized Meteorological Centre (RSMC) Tokyo-Typhoon Center of the Japan Meteorological Agency (JMA). Typhoons that lasted more than five days were included in this study, yielding over 300 tracks during the 1998–2017 period (Figure 1).

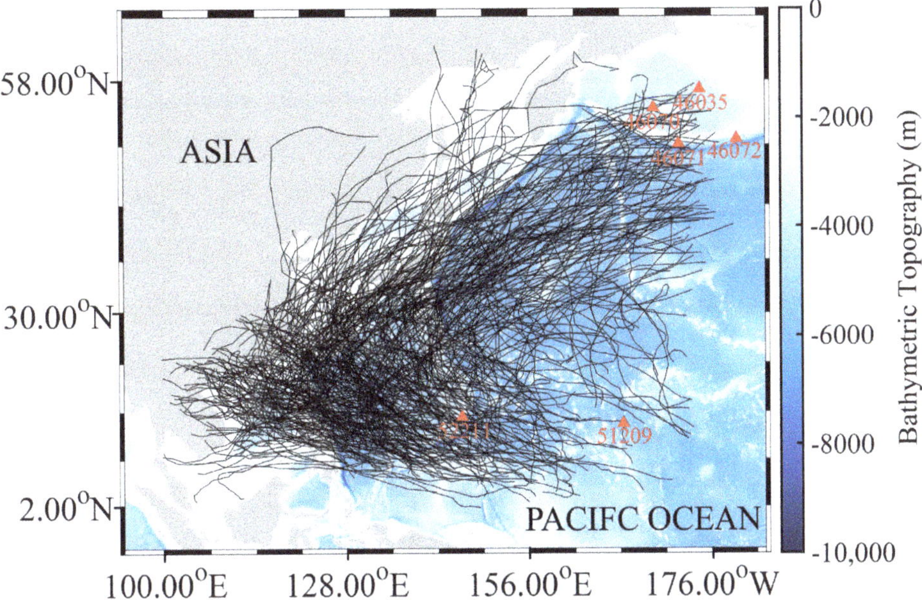

Figure 1. Bathymetric topography of the North West Pacific Ocean with high spatial resolution at a 0.1° grid in the area between 5° S and 65° N, 93° and 193° E. The color indicates the bathymetric topography. The red triangle indicates the locations of the National Data Buoy Center (NDBC) buoys. The black lines represent the typhoon track from Japan Meteorological Agency (JMA).

The WW3 model was developed by the National Centers for Environmental Prediction of the National Oceanic and Atmospheric Administration (NOAA NCEP). The WW3 model has been shown to perform well at simulating the characteristics of TC-induced waves [36,37]. In this study, 0.125° gridded ECMWF wind data provide forcing for WW3. Bathymetry data for the WW3 simulation were obtained at 0.1° horizontal resolution from GEBCO [38]. The resolution of the bathymetry data was 0.5°. The model was initialized at 00:00 UTC 1 January 1998 and terminated at 00:00 UTC 1 January 2018. Wave data were output at 0.5° spatial resolution and 30 min temporal resolution. Data from the six NDBC buoys allowed us to verify the WW3 results against more than three million observations during the simulation period. Figure 2 shows that the WW3-simulated results were close to the measurements from NDBC buoys (>300,000), showing a 0.43 m root mean square error (RMSE) with a 0.94 correlation (COR). The results indicate the reliability of the model.

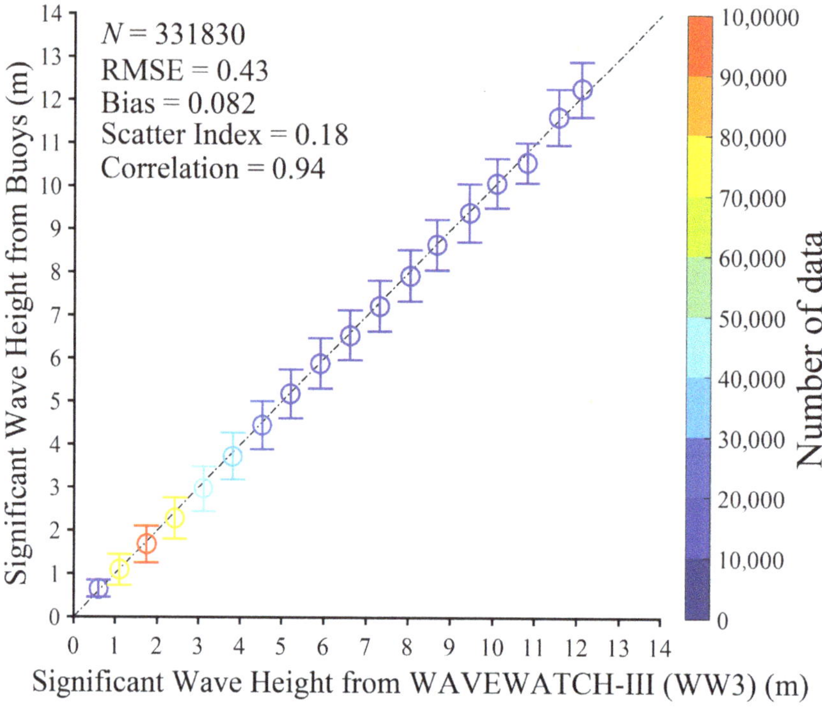

Figure 2. Comparison from WAVEWATCH-III (WW3)-simulated significant wave height (SWH) with the measurements from NDBC buoys, in which the error-bars represent the standard deviation of each bin at a 0.5 bin from 0 to 14 m of SWH.

3. Results

3.1. Classification of Typhoon Tracks

In the WNP, most typhoons either recurve or move approximately linearly, while just a few storms follow complex tracks such as loops. In this study, we only included typhoon tracks with durations of more than five days and omitted the small number of complex tracks. Using the track data, we clustered typhoon tracks according to genesis and extinction locations. If the genesis location is at the maximum longitude (farthest east) anywhere on the TC track, and the extinction location is at the minimum longitude (farthest west), the track is assigned to the Northeast Shift group. If the genesis location is located at the maximum longitude anywhere in the track (farthest east), and extinction location is at the minimum longitude (farthest west), the track is assigned to the straight-moving group. Within the straight-moving group, the average slope k of tropical cyclone track calculating between the genesis and extinction locations was used to further partition tracks. If the arctangent of k was between $7\pi/8$ and $9\pi/8$, the track was assigned to the west shift group. If the arctangent of k was between $5\pi/8$ and $7\pi/8$, the track was assigned to the northwest shift group [1,2,39]. If neither the genesis location nor the extinction location was located at the minimum longitude, the track belonged to the recurve group. Members of the recurve group were further partitioned by their westernmost location, known as the recurve point. If the longitude of the recurve point was >140°, the track was assigned to the east turn group. If the longitude of the recurve point <125°, the track was assigned to the west turn group. If the longitude of the recurve point was between 125 and 140°, the track was assigned to the middle turn group [40]. Thus, all typhoon tracks with duration >5 days were assigned to one of six track groups: northeast shift, east turn, middle turn, west turn, northwest shift, and west shift. The middle turn group was the most populous track group (*n* = 91), whereas the northeast shift was the least populous (*n* = 28; Figure 3).

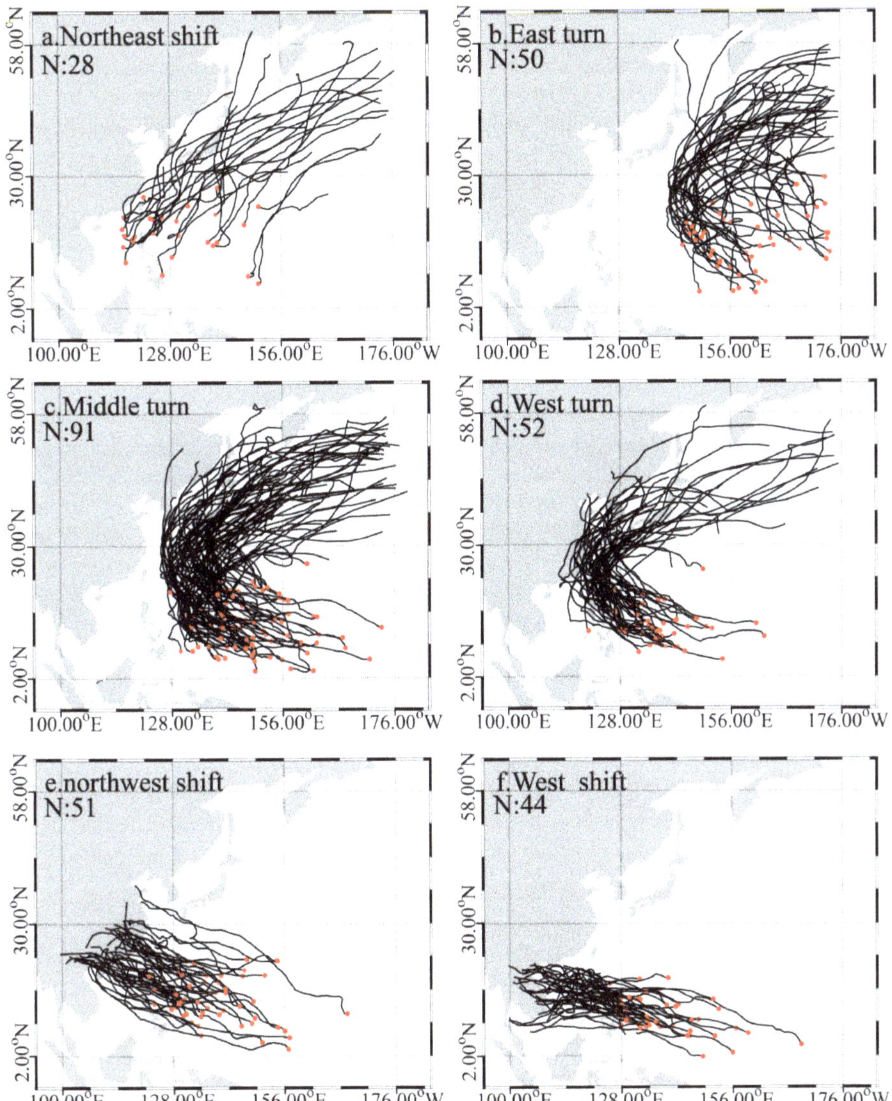

Figure 3. The six types of collected typhoon tracks from the JMA track data, in which the red spots represent the typhoons. (**a**) Northeast shift; (**b**) east turn; (**c**) middle turn; (**d**) west turn; (**e**) northwest shift; and (**f**) west shift.

3.2. Parameters of Typhoon-Induced Waves

In order to remove the background wave parameters without typhoons, we chose the wave parameters simulation result during the period of each typhoon path group. Figure 4 shows the average WW3-simulated SWHs from each typhoon path group, in which the black polylines represent the typical path in each path group. Maximum SWH (MS) values were at similar locations for all typhoon path groups: between 50° N–60° N and 160° E–187° W at the end of the typhoon's lifecycle. Furthermore, the asymmetrical structure was found to the right and left of the typhoon tracks. It was not surprising that this SWH pattern followed the tracks of the northeast shift, east turn, and middle turn groups (Figure 4a–c corresponding to Figure 3a–c), because no land barriers occurred in those regions to interrupt the wave pattern. In the more westward typhoon track groups (Figure 3d–f), two relatively high SWH regions were observed around 30° N (Figure 4c–f). In particular, the second SWH peak moved from the East China Sea to the South China Sea as typhoon tracks shifted more westward. Since West Shift typhoon tracks remained far from the northerly-located Monsoon region, waves in that area are determined by weather systems in other regions, such as the monsoon wind and the Westerly Belt wind at the northern Central Pacific Ocean [41]. The average mean wave period for each typhoon path group is shown in Figure 5; MWP patterns were also consistent with the typhoon track groups, especially for the northeast shift (Figure 5a) and northwest shift (Figure 5e) groups. The MWP peak area was located in the Japan Eastern Ocean in Figure 5a and the East China Sea in Figure 5e. However, maps of the average mean wave length (Figure 6) did not show this pattern. Taking the results together, the typhoon track appeared to substantially influence the basin-scale typhoon-induced wave distribution.

To further study the relationship between typhoon track and SWH distribution, we partitioned SWH values into individual wind sea and swell components by the WW3 model output. Figures 7 and 8 show the average wind sea and swell components, respectively, of simulated SWH for each typhoon track group. The wind sea SWH component reached maximum values of approximately 3 m near the typhoon tracks, while the swell SWH component reached values up to 2 m. Although the waves produced by other weather systems blur the signal of typhoon-induced waves at latitudes above 50° N, it is clear from Figure 7d–f that the wind sea SWH component varied coherently between the track groups. The maximum wind-sea SWH was on the typhoon path, such as the middle turn (Figure 7c), west turn (Figure 7d), and northwest shift (Figure 7e). However, no consistent relationship was observed between the swell SWH component and typhoon tracks. We therefore concluded that variations in typhoon track between the typhoon track groups primarily affected the distribution of SWH through the wind sea component of the wave system.

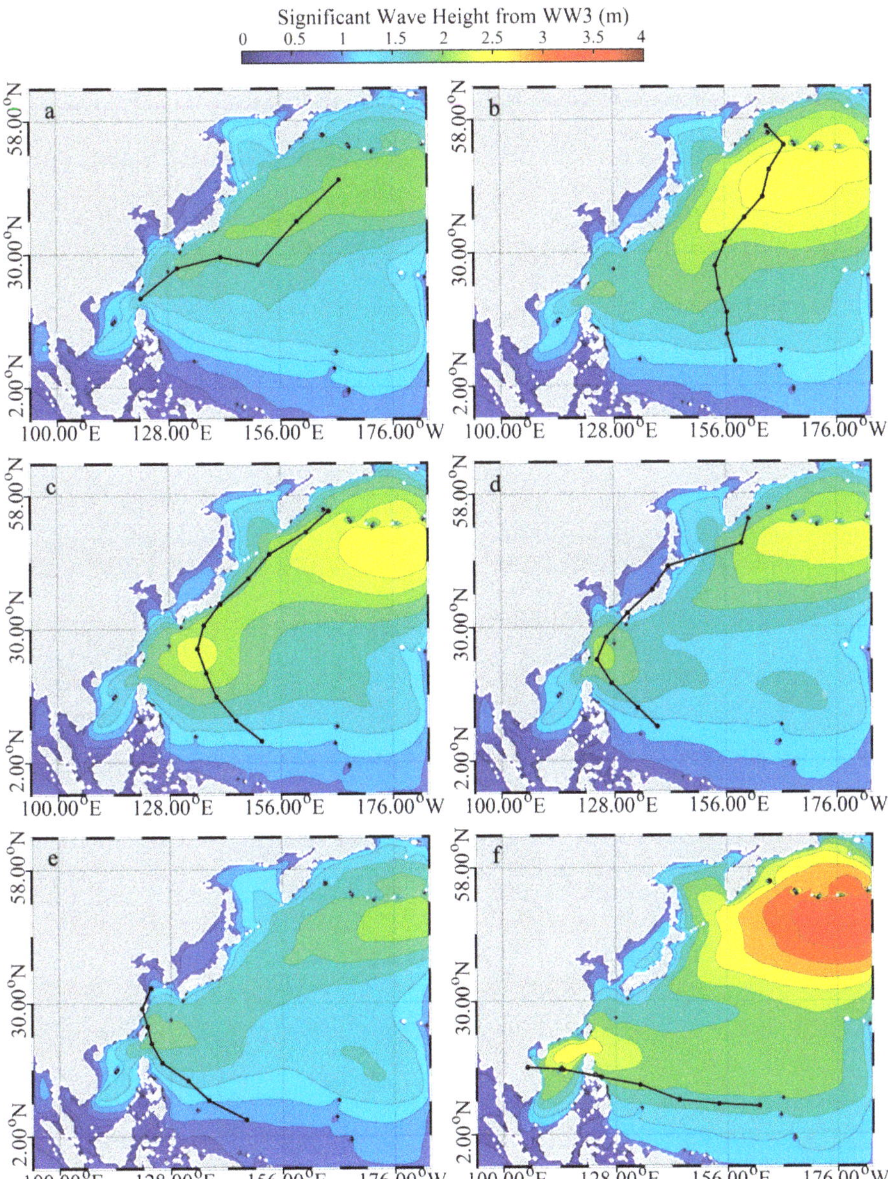

Figure 4. The average result of SWH from the WW3 for each type of typhoon paths from 1998 to 2018, in which the black polylines represent the typical paths for each type of typhoon paths. (**a**) Northeast shift; (**b**) east turn; (**c**) middle turn; (**d**) west turn; (**e**) northwest shift; and (**f**) west shift.

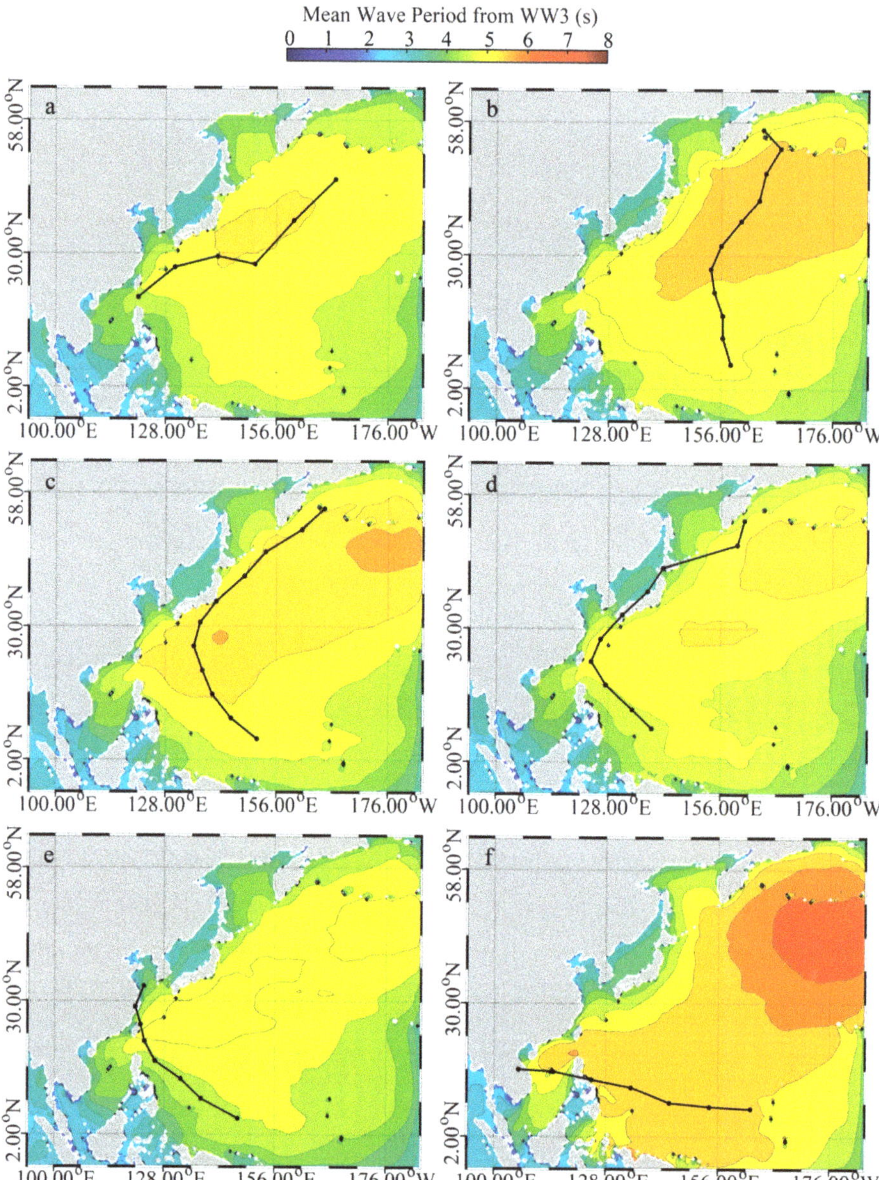

Figure 5. The average result of mean wave period (MWP) from the WW3 for each type of typhoon paths from 1998 to 2018, in which the black polylines represent the typical paths for each type of typhoon paths. (**a**) Northeast shift; (**b**) east turn; (**c**) middle turn; (**d**) west turn; (**e**) northwest shift; and (**f**) west shift.

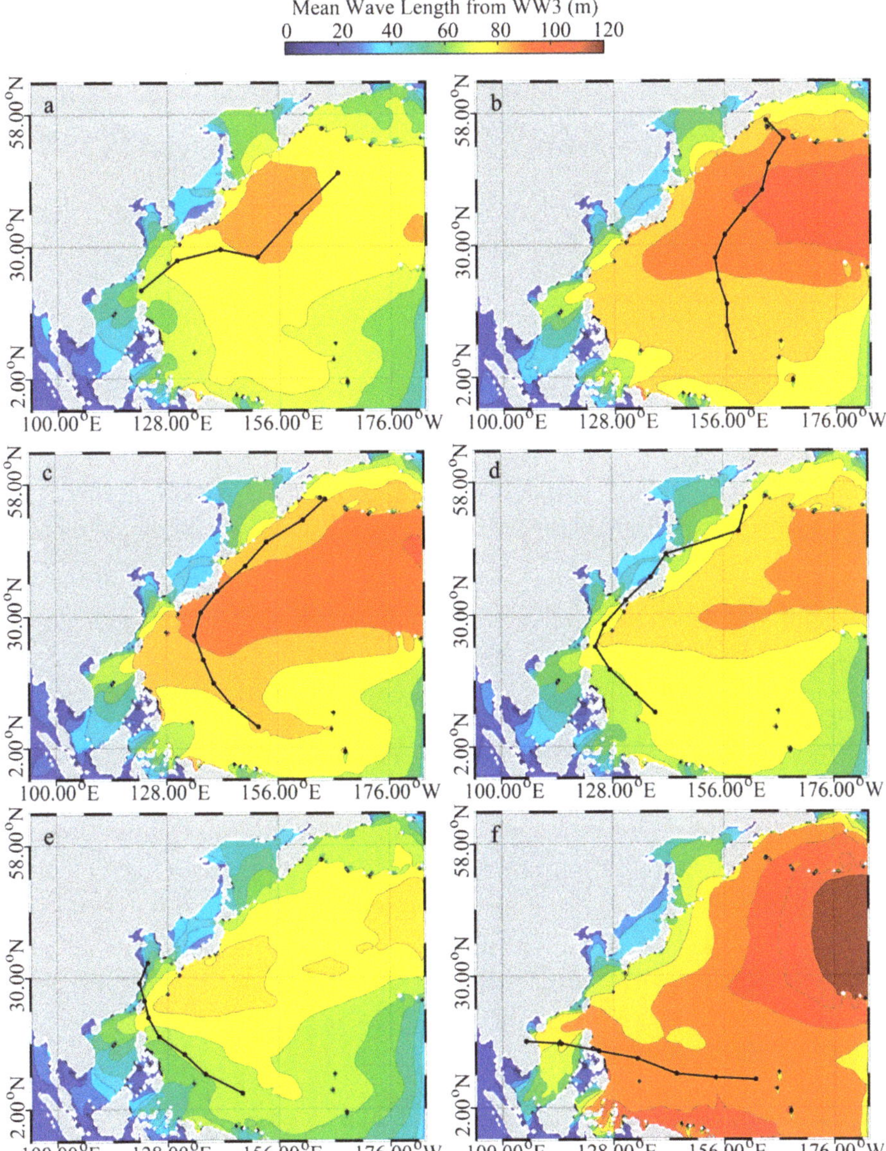

Figure 6. The average result of MWL from the WW3 for each type of typhoon paths from 1998 to 2018, in which the black polylines represent the typical paths for each type of typhoon paths. (**a**) Northeast shift; (**b**) east turn; (**c**) middle turn; (**d**) west turn; (**e**) northwest shift; and (**f**) west shift.

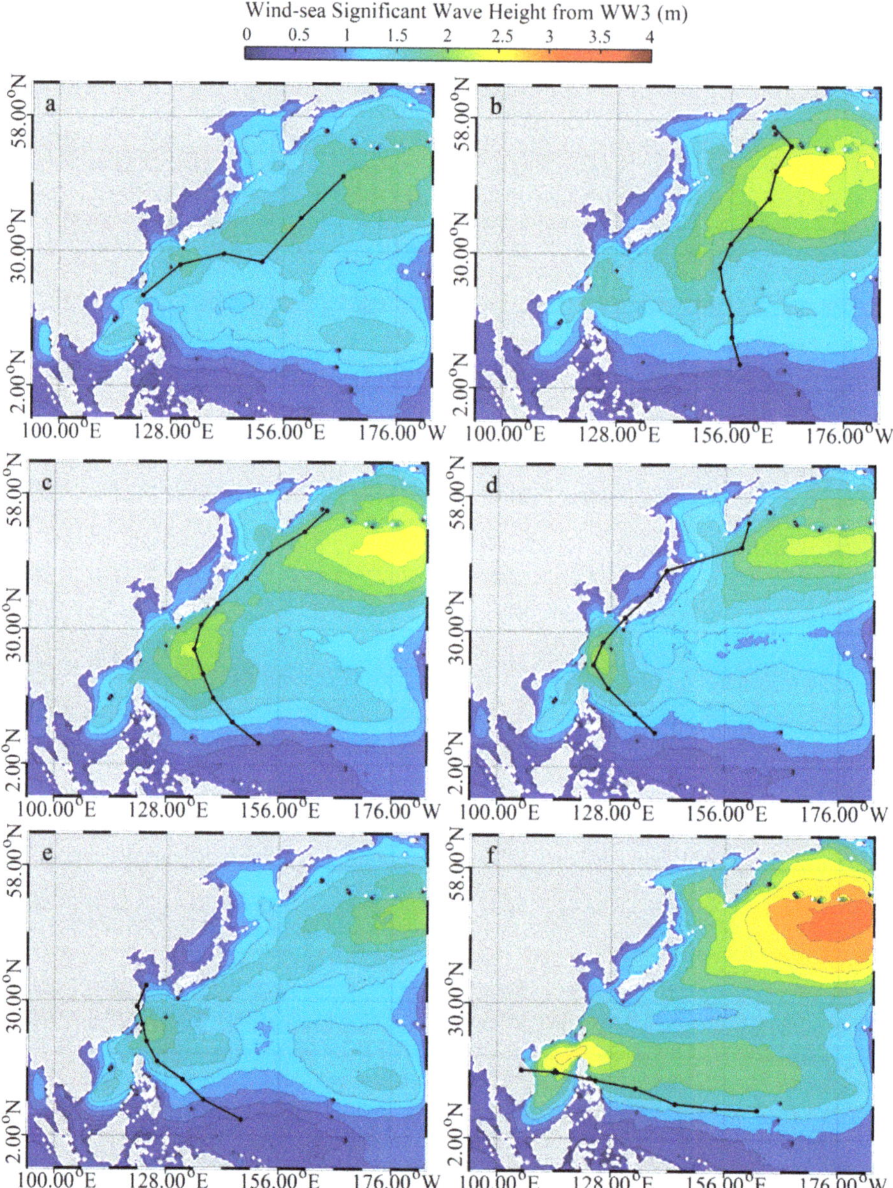

Figure 7. The average result of wind-sea SWH from the WW3 for each type of typhoon paths from 1998 to 2018, in which the black polylines represent the typical paths for each type of typhoon paths. (**a**) Northeast shift; (**b**) east turn; (**c**) middle turn; (**d**) west turn; (**e**) northwest shift; and (**f**) west shift.

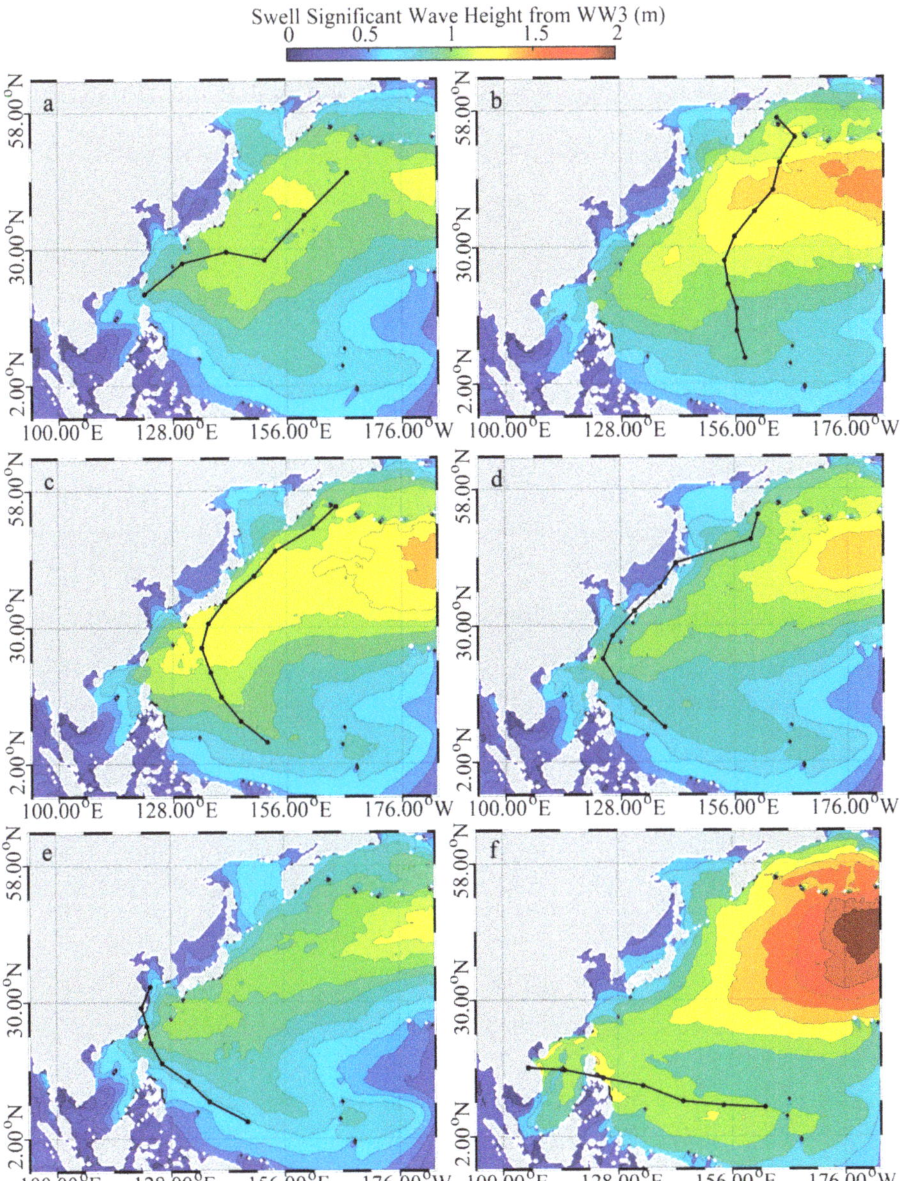

Figure 8. The average result of swell SWH from the WW3 for each type of typhoon paths from 1998 to 2018, in which the black polylines represent the typical paths for each type of typhoon paths. (**a**) Northeast shift; (**b**) east turn; (**c**) middle turn; (**d**) west turn; (**e**) northwest shift; and (**f**) west shift.

3.3. Analysis of the Leading EOF Modes of Significant Wave Height Variation

As for the EOF analysis for the SWH data on the typhoon path period, we did not find the consequent EOF modes result and the time series result remained interceptive. Thus, to study the spatial and temporal distribution of typhoon-induced SWH in the 1998–2017 period, empirical orthogonal function (EOF) analysis of average SWH was performed. EOF analysis decomposed the dataset into a series of orthogonal functions. The eigenvalue error method proposed by North et al. (1982) was applied to test the significance of the EOF outputs [42]. The range of eigenvalue error was calculated by the following equation.

$$e_j = \lambda_j \left(\frac{2}{n}\right)^{\frac{1}{2}} \tag{1}$$

in which, e_j is the eigenvalues error, λ_j is the eigenvalue, and n is the sample number. The judgment equation is stated as:

$$\lambda_j - \lambda_{j+1} > e_j \tag{2}$$

If the Equation (2) is derived, the two eigenvalues representing the modes of EOF is reliable.

As shown the Table 1, we found the four leading EOF modes of average SWH passed the significance test. The leading EOFs indicate the patterns that explain the greatest variation within the dataset.

Table 1. The information about the eigenvalue error method.

Modes	Eigenvalues (λ)	e_j	$\lambda_j - \lambda_{j+1}$
1	450.7115	7.4577	230.7340
2	219.9775	3.6398	36.5925
3	183.3850	3.0344	24.3827
4	159.0023	2.6309	26.6687

The spatial distributions of the four leading EOF modes of average SWH are shown in Figure 9. From the EOF patterns we computed two principal component variables: PC(t), which represents the amplitude of the EOF pattern at a single time, and PC(max), which represents the maximum EOF pattern amplitude during the 1998–2017 period. The first EOF mode (Figure 9a) contributes 17.3% of the total variance in average SWH. It represents a dipole in average SWH between the central North Pacific and the Western Pacific near Japan. This indicates that the typhoon-induced wave energy dominated near Japan, as well as in the Sea of Japan and the East China Sea. The second mode (Figure 9b) contributed to 8.4% of the total variance. This EOF shows a tripole pattern, with elevated SWH values south of Japan and Korea associated with depressed values south of the Kamchatka Peninsula, and elevated values near the Aleutian Islands. The third (Figure 9c) and fourth (Figure 9d) EOF modes contributed to 8.4% and 7.0% of the total variance, respectively. The corresponding time series of PC(t)/PC(max) are presented in Figure 10, in which the values were normalized into a range between −0.5 and 0.5. The amplitude of PC(t)/PC(max) was clearly larger for the first EOF mode (Figure 10a) than for the other three modes. The first EOF mode oscillates on a roughly 1-year cycle, varying between approximately 0.4 and −0.5. This behavior indicates an interannual variability pattern in typhoon-induced SWH distribution over the Western North Pacific. For the second through fourth EOF modes (Figure 10b–d), the time series of PC(t)/PC(max) show no clear cyclical trends.

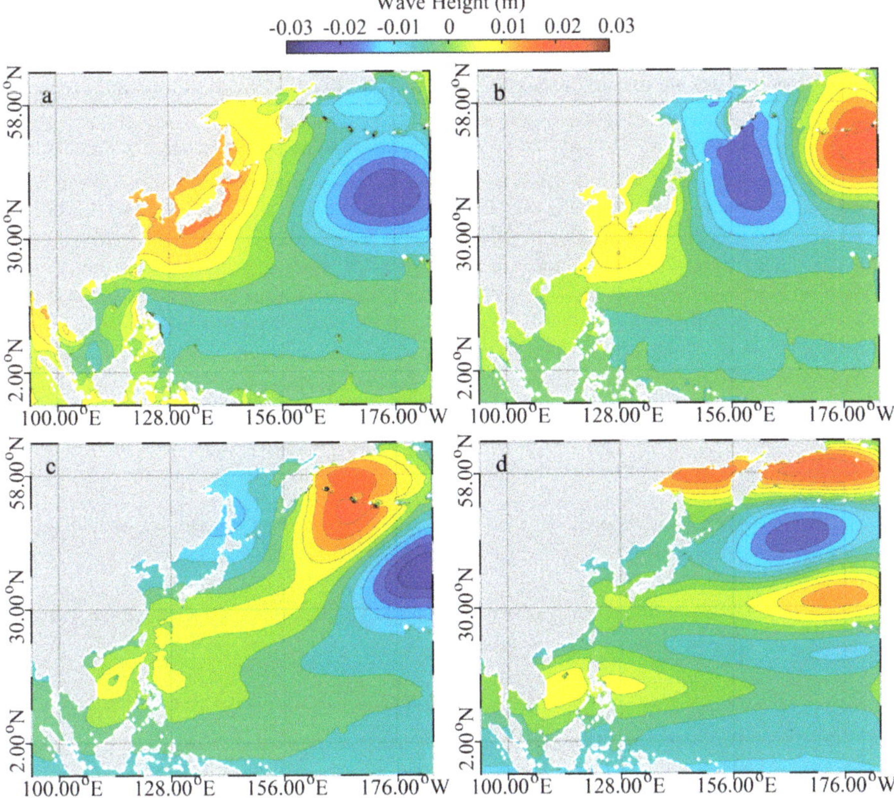

Figure 9. Spatial distribution of the four dominant EOF modes for average SWH from 1998 to 2018. (**a**) The first mode; (**b**) the second mode; (**c**) the third mode; and (**d**) the forth mode.

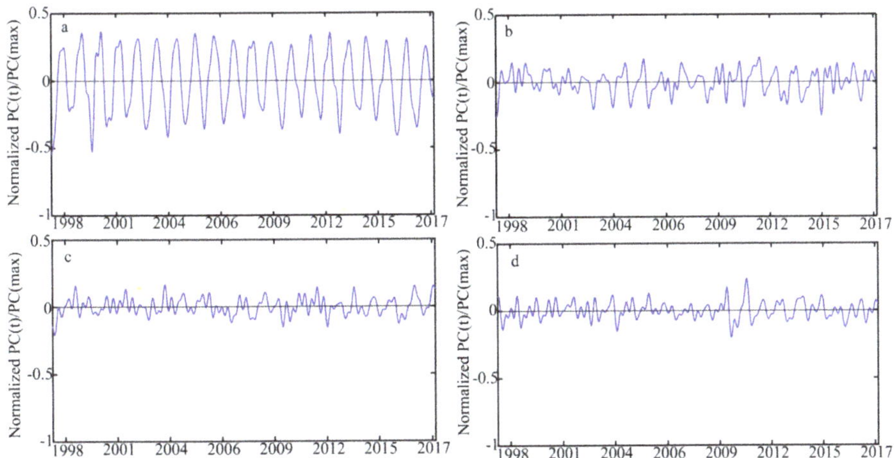

Figure 10. Time series of the amplitudes of the four dominant EOF modes for average SWH from 1998 to 2018. (**a**) The first mode; (**b**) the second mode; (**c**) the third mode; and (**d**) the forth mode.

4. Discussion

As for the EOF modes result (Figure 9), we found the EOF modes distribution in the East China Sea was not completely consistent with the result in Wang et al. (2016) [43]. Although a similar magnitude was achieved in the first mode, the big rate was located in the North East China Sea and the trend of EOF, which is different from the result in [43]. There was a high wave in the Japan Sea caused by the strong seasonal northerly wind. In the Pacific Ocean, summer monsoon seasonal wind (south westerly) and typhoons may cause a high sea state. Such seasonality was derived by the EOF analysis in Figure 9a and this behavior is also proposed in [43]. In the second and third modes, there was no consistency with two analyses. As for the corresponding time series of PC(t)/PC(max) (Figure 10), our analysis had a similar trend of the time period result in [35,43].

Here, we analyzed the interannual variability of PC(t)/PC(max) for all typhoons from 1998 to 2017. Figure 11 shows the values of PC(t)/PC(max) of the first EOF for each typhoon track group. Overall, PC(t)/PC(max) values were positive, indicating that typhoons tended to produce higher SWH values near Japan and in nearby seas, and lower values over the central North Pacific (Figure 9a). However, in the West Shift group (Figure 11f), the PC(t)/PC(max) values generally decreased in amplitude, indicating a weakening signal from westward-tracking typhoons during the 20-year period studied. In the northeast shift (Figure 11a), west turn (Figure 11d), and northwest shift groups, the PC(t)/PC(max) values remained relatively constant, indicating that no climate trend in the typhoon-induced waves pattern could be detected from 1998 to 2017. In the east turn (Figure 11b) and middle turn (Figure 11c) track groups, PC(t)/PC(max) values fluctuated substantially during the 1998–2017 period. Indeed, it is better to study the change in the climate of typhoon-induced wave distribution with a minimum 30-year period. Nevertheless, we presumed that this short-term behavior might reflect possible changes in the climate of typhoon-induced wave distribution.

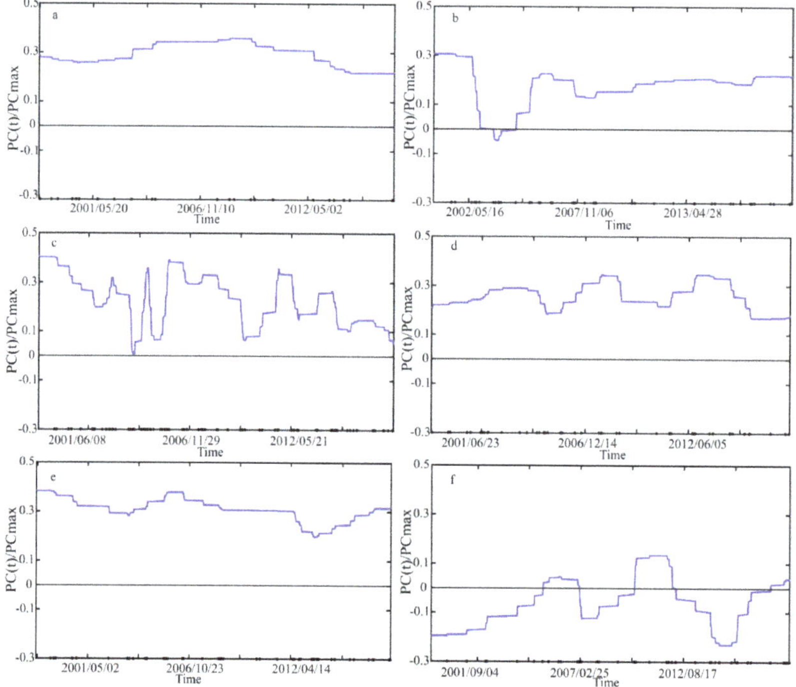

Figure 11. The first mode principal components (PCs) part results in each path period. (**a**) Northeast shift; (**b**) east turn; (**c**) middle turn; (**d**) west turn; (**e**) northwest shift; and (**f**) west shift.

5. Conclusions

The climate of tropical cyclone-induced waves remains an understudied topic in the oceanography community. In this study, we analyzed the wave distribution from typhoons in the Western North Pacific Ocean during the period 1998–2017. Typhoon tracks were partitioned into six groups based on track direction and longitude of recurvature. The WAVEWATCH-III Model (WW3; Version 5.16) was used to simulate waves over the Western North Pacific from 1998 to 2017. Simulated wave parameters (significant wave height (SWH), mean wave period (MWP), and mean wave length (MWL)) associated with each typhoon track group were analyzed. Simulation results revealed that the typhoon track substantially affected typhoon-induced wave distribution, especially for SWH and MWP. Across the track groups, elevated SWH values tended to occur in a symmetrical structure cantered on the typhoon path. Typhoons that recurved between 125° E and 140° E, or west of 140° E, produced a region of elevated SWH values south and southwest of the Japanese main islands, respectively, with minimum SWH values near 30° N. In the farther westward typhoon track groups (west turn, northwest shift, and west shift), two relatively high SWH regions were observed around 30° N. In particular, the secondary peak of SWH moved from the East China Sea to the South China Sea when typhoon tracks shifted farther westward. Furthermore, partitioning simulated SWH into wind sea and swell components indicates that typhoon track differences primarily contributed to differences in the wind sea component of SWH.

Empirical orthogonal function (EOF) analysis was used to evaluate the main modes of SWH variation across the set of simulated typhoon-induced waves. The first mode contributed to 17.3% of the total variance, with other modes contributing less than 10%. The first EOF mode was associated with a dipole of SWH values, with opposite values near Japan and in the central North Pacific.

The time evolution of the principal component (PC) of EOF 1 for each typhoon track group, relative to the maximum PC value during 1998–2017, was calculated to further study the climate of WW3-simulated SWH for each type of typhoon path. The first EOF mode revealed that the SWH pattern oscillated on an approximately 1-year cycle and thus appeared to show possible short-term climate signals within the 1998–2017 period, while other EOF modes had no clear cyclical trends. Typhoons that recurved east of 140° E, and those that tracked westward toward Southeast Asia, oscillated substantially during the period, and thus appeared to show short-term natural variability signals within the 1998–2017 period. In the near future, we planned to perform more analyses about the simulated waves using the WW3 model under the background of typhoons, which remove the basic characteristics of ocean waves in the Northwestern Pacific.

Author Contributions: W.S., Y.W. and J.Z. came up the original idea and designed the experiments. Y.H. and J.Z. contribute to wave simulations from the WW3 model. W.S. and Y.H. analyzed the dataset. Y.W. and J.Z. provided great help for the data analysis and discussions. All authors contributed to the writing and revising of the manuscript. All authors have read and agreed to the published version of the manuscript.

Funding: This research was funded by the National Key Research and Development Program of China, grant number 2017YFA0604901, National Natural Science Foundation of China, grant number 41806005, 41776183 and 41976174, Public Welfare Technical Applied Research Project of Zhejiang Province of China, grant number LGF19D060003 and Science and Technology Project of Zhoushan City of China, grant number 2019C21008.

Acknowledgments: We appreciate the National Centers for Environmental Prediction (NCEP) of the National Oceanic and Atmospheric Administration (NOAA) for providing the source code for the WAVEWATCH-III (WW3) model. The European Centre for Medium-Range Weather Forecasts (ECMWF) wind data were accessed via http://www.ecmwf.int. The General Bathymetry Chart of the Oceans (GEBCO) data were downloaded via ftp.edcftp.cr.usgs.gov. The National Data Buoy Center (NDBC) provided buoy data via http://www.ndbc.noaa.gov/. Typhoon parameters were provided by the Japan Meteorological Agency (JMA) via http://www.jma.go.jp.

References

1. Yuan, J.P.; Jiang, J. The relationships between tropical cyclone tracks and local SST over the western north pacific. *J. Trop. Meteorol.* **2011**, *17*, 120–127.
2. Kim, H.K.; Seo, K.H. Cluster analysis of tropical cyclone tracks over the western north pacific using a self-organizing map. *J. Clim.* **2016**, *29*, 3731–3750. [CrossRef]
3. Akpınar, A.; Vledder, G.P.V.; Kömürcü, M.İ.; Özger, M. Evaluation of the numerical wave model (SWAN) for wave simulation in the Black Sea. *Cont. Shelf Res.* **2012**, *50*, 80–99. [CrossRef]
4. Kim, T.R.; Lee, J.H. Comparison of high wave hindcasts during Typhoon Bolaven (1215) using SWAN and WAVEWATCH III model. *J. Coast. Res.* **2018**, *85*, 1096–1100. [CrossRef]
5. Bi, F.; Song, J.B.; Wu, K.J.; Xu, Y. Evaluation of the simulation capability of the Wavewatch III model for Pacific Ocean wave. *Acta Oceanol. Sin.* **2015**, *34*, 43–57. [CrossRef]
6. Zheng, K.W.; Sun, J.; Guan, C.L.; Shao, W.Z. Analysis of the global swell and wind sea energy distribution using WAVEWATCH III. *Adv. Meteorol.* **2016**, *2016*, 8419580. [CrossRef]
7. Shukla, R.P.; Kinter, J.L.; Shin, C.S. Sub-seasonal prediction of significant wave heights over the Western Pacific and Indian Oceans, part II: The impact of ENSO and MJO. *Ocean Model.* **2018**, *123*, 1–15. [CrossRef]
8. He, H.L.; Xu, Y. Wind-wave hindcast in the Yellow Sea and the Bohai Sea from the year 1988 to 2002. *Acta Oceanol. Sin.* **2016**, *35*, 46–53. [CrossRef]
9. Zheng, K.W.; Osinowo, A.A.; Sun, J.; Hu, W. Long-term characterization of sea conditions in the East China Sea using significant wave height and wind speed. *J. Ocean Univ. China* **2018**, *17*, 733–743. [CrossRef]
10. Gallagher, S.; Tiron, R.; Dias, F. A long-term nearshore wave hindcast for Ireland: Atlantic and Irish Sea coasts (1979–2012). *Ocean Dyn.* **2014**, *64*, 1163–1180. [CrossRef]
11. Guo, L.L.; Perrie, W.; Long, Z.X.; Toulany, B.; Sheng, J.Y. The impacts of climate change on the autumn North Atlantic wave climate. *Atmosphere-Ocean* **2015**, *53*, 491–509. [CrossRef]
12. Shao, W.Z.; Sheng, Y.X.; Li, H.; Shi, J.; Ji, Q.Y.; Tai, W.; Zuo, J.C. Analysis of wave distribution simulated by WAVEWATCH-III model in typhoons passing Beibu Gulf, China. *Atmosphere* **2018**, *8*, 265. [CrossRef]
13. Sheng, Y.X.; Shao, W.Z.; Li, S.Q.; Zhang, Y.M.; Yang, H.W.; Zuo, J.C. Evaluation of typhoon waves simulated by WaveWatch-III model in shallow waters around Zhoushan islands. *J. Ocean Univ. China* **2019**, *19*, 365–375. [CrossRef]
14. Fan, Y.; Lin, S.J.; Held, I.M.; Yu, Z.; Tolman, H.L. Global ocean surface wave simulation using a coupled atmosphere-wave model. *J. Clim.* **2012**, *25*, 6233–6252. [CrossRef]
15. Gallagher, S.; Gleeson, E.; Tiron, R.; Mcgrath, R.; Dias, F. Wave climate projections for Ireland for the end of the 21st century including analysis of EC-Earth winds over the North Atlantic Ocean. *Int. J. Clim.* **2016**, *36*, 4592–4607. [CrossRef]
16. Li, S.Q.; Guan, S.D.; Hou, Y.J.; Liu, Y.H.; Bi, F. Evaluation and adjustment of altimeter measurement and numerical hindcast in wave height trend estimation in China's coastal seas. *Int. J. Appl. Earth Obs. Geoinf.* **2018**, *67*, 161–172. [CrossRef]
17. Li, X.F. The first Sentinel-1 SAR image of a typhoon. *Acta Oceanol. Sin.* **2015**, *34*, 1–2. [CrossRef]
18. Li, X.F.; Zhang, J.A.; Yang, X.F.; Pichel, W.G.; DeMaria, M.; Long, D.; Li, Z.W. Tropical cyclone morphology from spaceborne synthetic aperture radar. *Bull. Am. Meteorol. Soc.* **2013**, *94*, 215–230. [CrossRef]
19. Ji, Q.Y.; Shao, W.Z.; Sheng, Y.X.; Yuan, X.Z.; Sun, J.; Zhou, W.; Zuo, J.C. A promising method of cyclone wave retrieval from Gaofen-3 synthetic aperture radar image in VV-polarization. *Sensors* **2018**, *18*, 2064. [CrossRef] [PubMed]
20. Romeiser, R.; Graber, H.C.; Caruso, M.J.; Jensen, R.E.; Walker, D.T.; Cox, A.T. A new approach to ocean wave parameter estimates from C-band ScanSAR images. *IEEE Trans. Geosci. Remote Sens.* **2015**, *53*, 1320–1345. [CrossRef]
21. Shao, W.Z.; Li, X.F.; Hwang, P.A.; Zhang, B.; Yang, X.F. Bridging the gap between cyclone wind and wave by C-band SAR measurements. *J. Geophys. Res. Oceans* **2017**, *122*, 6714–6724. [CrossRef]
22. Shao, W.Z.; Hu, Y.Y.; Yang, J.S.; Nunziata, F.; Sun, J.; Li, H.; Zuo, J.C. An empirical algorithm to retrieve significant wave height from Sentinel-1 synthetic aperture radar imagery collected under cyclonic conditions. *Remote Sens.* **2018**, *10*, 1367. [CrossRef]

23. Ding, Y.Y.; Zuo, J.C.; Shao, W.Z.; Shi, J.; Yuan, X.Z.; Sun, J.; Hu, J.C.; Li, X.F. Wave parameters retrieval for dual-polarization C-band synthetic aperture radar using a theoretical-based algorithm under cyclonic conditions. *Acta Oceanol. Sin.* **2019**, *38*, 21–31. [CrossRef]

24. Chen, Y.P.; Xie, D.M.; Zhang, C.K.; Qian, X.S. Estimation of long-term wave statistics in the East China Sea. *J. Coast. Res.* **2013**, *65*, 177–182. [CrossRef]

25. Hwang, P.A.; Bratos, S.M.; Teague, W.J.; Wang, D.W.; Jacobs, G.A.; Resio, D.T. Winds and waves in the yellow and east China seas: A comparison of space-borne altimeter measurements and model results. *J. Oceanogr.* **1999**, *55*, 307–325. [CrossRef]

26. Liang, B.C.; Liu, X.; Li, H.J.; Wu, Y.J.; Lee, D.Y. Wave Climate Hindcasts for the Bohai Sea, Yellow Sea, and East China Sea. *J. Coast. Res.* **2014**, *32*, 172–180.

27. Markina, M.Y.; Gavrikov, A.V. Wave climate variability in the North Atlantic in recent decades in the winter period using numerical modeling. *Oceanology* **2016**, *56*, 320–325. [CrossRef]

28. Divinsky, B.V.; Kosyan, R.D. Spatiotemporal variability of the Black Sea wave climate in the last 37 years. *Cont. Shelf Res.* **2017**, *136*, 1–19. [CrossRef]

29. Caires, S.; Swail, V.R.; Wang, X.L.L. Projection and Analysis of Extreme Wave Climate. *J. Clim.* **2006**, *19*, 5581–5605. [CrossRef]

30. Li, J.X.; Chen, Y.Q.; Pan, S.Q. Modelling of Extreme Wave Climate in China Seas. *J. Coast. Res.* **2016**, *75*, 522–526. [CrossRef]

31. He, H.L.; Song, J.B.; Bai, Y.F.; Xu, Y.; Wang, J.J.; Bi, F. Climate and extrema of ocean waves in the East China Sea. *Sci. China Earth Sci.* **2018**, *7*, 1–15. [CrossRef]

32. Sartini, L.; Mentaschi, L.; Besio, G. Comparing different extreme wave analysis models for wave climate assessment along the Italian coast. *Coast. Eng.* **2015**, *100*, 37–47. [CrossRef]

33. Suursaar, Ü.; Kullas, T. Decadal changes in wave climate and sea level regime: The main causes of the recent intensification of coastal geomorphic processes along the coasts of Western Estonia? *WIT Trans. Ecol. Environ.* **2009**, *126*, 105–116.

34. Shimura, T.; Mori, N.; Mase, H. Future Projection of Ocean Wave Climate: Analysis of SST Impacts on Wave Climate Changes in the Western North Pacific. *J. Clim.* **2015**, *28*, 3171–3190. [CrossRef]

35. Niroomandi, A.; Maa, G.F.; Ye, X.Y.; Lou, S.; Xue, P.F. Extreme value analysis of wave climate in Chesapeake Bay. *Ocean Eng.* **2018**, *159*, 22–36. [CrossRef]

36. Liu, Q.X.; Babanin, A.; Fan, Y.L.; Zieger, S.; Guan, C.L.; Moon, I.J. Numerical simulations of ocean surface waves under hurricane conditions: Assessment of existing model performance. *Ocean Model.* **2017**, *118*, 73–93. [CrossRef]

37. Zieger, S.; Babanin, A.V.; Rogers, W.E.; Young, I.R. Observation-based source terms in the third-generation wave model WAVEWATCH. *Ocean Model.* **2015**, *96*, 2–25. [CrossRef]

38. Weatherall, P.; Marks, K.M.; Jakobsson, M.; Schmitt, T.; Tani, S.; Arndt, J.E.; Rovere, M.; Chayes, D.; Ferrini, V.; Wigley, R. A new digital bathymetric model of the world's oceans. *Earth Space Sci.* **2015**, *2*, 331–345. [CrossRef]

39. Nakamura, J.; Lall, U.; Kushnir, Y.; Camargo, S.J. Classifying North Atlantic Tropical Cyclone Tracks by Mass Moments. *J. Clim.* **2009**, *22*, 5481–5494. [CrossRef]

40. Yin, H.; Wang, Y.Q.; Zhong, W. Characteristics and influence factors of the rapid intensification of tropical cyclone with different tracks in Northwest Pacific. *J. Meteorol. Sci.* **2016**, *36*, 194–203. (In Chinese)

41. Zhu, Z.X.; Zhou, K. Simulation and Analysis of Ocean Wave in the Northwest Pacific Ocean. *Appl. Mech. Mater.* **2012**, *212*, 430–435. [CrossRef]

42. North, G.R.; Bell, T.L.; Cahalan, R.F.; Moeng, F.J. Sampling Errors in the Estimation of Empirical Orthogonal Functions. *Mon. Weather Rev.* **1982**, *110*, 699. [CrossRef]

43. Wang, J.; Dong, C.M.; He, Y.J. Wave climatological analysis in the East China Sea. *Cont. Shelf Res.* **2016**, *120*, 26–40. [CrossRef]

Article

Consequences of a Storm Surge for Aeolian Sand Transport on a Low-Gradient Beach

Jorn T. Tuijnman [†], Jasper J. A. Donker [‡], Christian S. Schwarz [§] and Gerben Ruessink [*]

Department of Physical Geography, Faculty of Geosciences, Utrecht University, P.O. Box 80.115, 3508 TC Utrecht, The Netherlands; jornt.tuijnman@gmail.com (J.T.T.); jasperdonker@gmail.com (J.J.A.D.); cschwarz@udel.edu (C.S.S.)

* Correspondence: b.g.ruessink@uu.nl; Tel.: +31-30-2532780
† Current address: Aqua Vision, Servaasbolwerk 11, 3512 NK Utrecht, The Netherlands.
‡ Current address: Geodan, President Kennedylaan 1, 1079 MB Amsterdam, The Netherlands.
§ Current address: School of Marine Science and Policy, College of Earth, Ocean & Environment, University of Delaware, Newark, DE 19716, USA.

Received: 2 July 2020; Accepted: 3 August 2020; Published: 5 August 2020

Abstract: Wind-blown beach sand is the primary source for the volume growth of the most seaward dune, the foredune. Strong wind events can potentially dominate long-term aeolian supply but in reality do not contribute considerably because they often coincide with a storm surge. The aim of this paper is to further our understanding of how a storm surge prevents or severely restricts aeolian supply. Using field data collected on the 1:50 sloping Egmond beach (Netherlands) in the aftermath of a 1-m storm surge, we show that the ground water in the upper beach rose to well above normal levels during the surge, which resulted in the development of a seepage face during falling tide and hence persistent saturation of the emerging beach. Using a fetch-based model, we predicted aeolian supply during the 2-day surge period to be about 66% of the potential supply. Fetch limitations imposed by the surge-induced inundation and the continuous saturation of the sand on the emerging beach both contributed to the predicted supply limitation. Our results quantitatively support earlier studies that suggested surges to be the primary condition that causes predictions of long-term potential foredune growth to overestimate measured growth.

Keywords: aeolian processes; surface moisture; storm surge; supply limitations; fetch

---

1. Introduction

Coastal foredunes are common morphological features along the world's sandy wave-dominated beaches and barrier systems [1,2]. Their dynamics reflect a sand-sharing system with the neighboring beach. During a single high-intensity storm or a cluster of storms a foredune can be eroded severely by marine processes [3–5], with the eroded sand being deposited on the beach and in the nearshore zone. The key hydrodynamic processes controlling dune erosion are reasonably well understood [4,6,7] and, accordingly, process-based models [8] can hindcast observed erosion events fairly well [5,9]. Foredune growth and recovery takes place through the deposition of wind-blown beach sand in vegetation and is a slow process in the sense that it may take seasons, years or longer before a previously eroded volume has been restored [10]. Accordingly, foredune growth is the cumulative effect of numerous aeolian transport events that differ in magnitude and duration [11]. A simple approach to compute the aeolian supply is the use of an equilibrium (or, potential) transport equation, in which the transport is a function of grain size and wind speed. Wind direction is additionally included to reflect that only onshore supply contributes to foredune growth [12,13]. Although the potential supply has been shown to match foredune volume growth reasonably well in some cases [14–16], it more often exceeds measured growth substantially [12,14,17–19].

Model failure may, at least in part, be due to inappropriate input wind data [19–21]. Near-surface wind speeds measured on land (but near the shore) are generally lower than those at sea [22,23]. Using offshore wind measurements without correcting for this systematic difference will thus overestimate potential transport rates and deposited volumes. The foredune will additionally alter the regionally representative (on land) wind speed and direction. Offshore directed winds may reverse on the upper beach and the seaward side of the foredune to cause onshore transport and foredune growth [20,24], while onshore directed winds are decelerated on the beach, causing wind speeds in the region where aeolian transport is initiated to be lower than the regionally representative wind speed [19,25]. In addition, onshore oblique winds will deflect in the alongshore direction at the foot of the foredune [19,20,26], resulting in alongshore rather than onshore aeolian transport. These topographic effects are strongest when the foredune is high and steep (e.g., scarped). For a Dutch case study with a 22-m high foredune with a 1:2.5 seaward slope, De Winter et al. [19] predicted the potential deposition volume to drop from 86 to 32 $m^3/m/yr$ when including correction factors for the foredune-induced topographical modification of the regionally representative onshore wind. Despite the marked reduction, the corrected potential deposition still exceeded the measured deposition of about 15 $m^3/m/yr$. This illustrates the common assertion that model failure is primarily due to the neglect of factors that reduce the aeolian supply to below its potential value. Surface moisture, an extensively studied supply-limiting factor, impedes the entrainment of sand [27], which can make part of the beach unavailable as a supply source [28,29]. Surface moisture may also increase the downwind distance needed to achieve the potential (maximum) transport rate (critical fetch [18,30]) to a value larger than the available beach-width. In such a case, the amount of sand blown across the landward margin of the beach onto the foredune is thus less than the potential amount. Model failure is indeed most pronounced on narrow beaches, with a width of about 200 to 300 m marking the upper value for which width rather than wind speed dominates the control on long-term aeolian supply to the foredune [14]. Lag deposits [31,32] and algae or salt crusts [33] are other examples of supply-limiting factors.

Data sets of aeolian activity on the beach spanning a wide range of conditions have illustrated that strong onshore wind events are most supply limited [11,12,34–36]. Such events are often associated with rainfall, high waves and hence substantial run-up on the beach and, in particular, elevated water levels because of wave-breaking induced set-up and a storm surge. The flooding of the beach will reduce beach-width and consequently reduce the transport at the beach-dune transition to well below the potential maximum. When the water levels are high enough, the result will not be deposition on the foredune but rather foredune scarping and erosion. Despite the fact that strong wind events are rather uncommon, the supply limitation may be so strong that they constitute the primary condition under which most of the over-prediction of foredune growth with a potential transport equation arises [11,34,35]. Nonetheless, the reasons how a surge partially or fully shuts down the aeolian system are not entirely understood. In addition to reducing beach-width, a surge also elevates the ground water level in the upper beach for several days [37] with potentially large but unexplored effects on spatio-temporal surface moisture dynamics and hence aeolian processes during and after the surge. Collecting data relevant to aeolian processes during a surge still poses a challenging task. Probably because of a lack of suitable data, recent model improvements to predict aeolian sand transport to the foredune in supply limited situations [18,38,39] have, with the exception of Cohn et al. [40], not been applied in detail to surge conditions. Cohn et al. [40]'s model indicated that run-up under certain conditions with elevated water levels can contribute to foredune growth rather than to scarping and erosion.

The overarching aim of this paper is to further our understanding of aeolian supply limitations imposed by a storm surge. To that end, we first analyze observations of cross-shore surface moisture dynamics under storm-surge conditions collected at Egmond beach, the Netherlands, during the Aeolex-II campaign in October 2017. The storm surge elevated the mean water level to about 1 m above astronomical tidal level, with the largest swash motions during the high-tide storm peak just reaching the beach-dune transition. We then quantify the surge-induced consequences for the aeolian supply to the foredune with the recently developed fetch-based model of Hage et al. [39] and examine the reasons for the predicted supply limitations (e.g., beach-width; moisture dynamics) with several model scenarios.

2. Methodology

2.1. Observations

2.1.1. Study Site

The field site was located approximately 3 km south of Egmond aan Zee, the Netherlands in the immediate vicinity of beach pole 41 (Figure 1). The coast at Egmond aan Zee is approximately North-South orientated (7° clockwise rotated with respect to North) and predominantly exposed to North-sea generated wind waves [5,41]. The intertidal beach at Egmond has a mild slope (typically, 1:40) and often contains a single slipface ridge [42], which are typical beach characteristics for this part of the Dutch coast [43,44]. At low tide, the beach is up to 100 m wide. The established foredune is 20–25 m high, has a steep (1:2.5) seaward-facing slope and, from a height of about 10 to 15 m above mean sea level (MSL), is densely covered in European marram grass (*Ammophila arenaria*) [45]. At a slightly larger height, the slope of the foredune lessens substantially toward the crest. This abrupt change in slope marks the location to which storms have previously eroded the foredune by means of scarping or rotational failure [5,46].

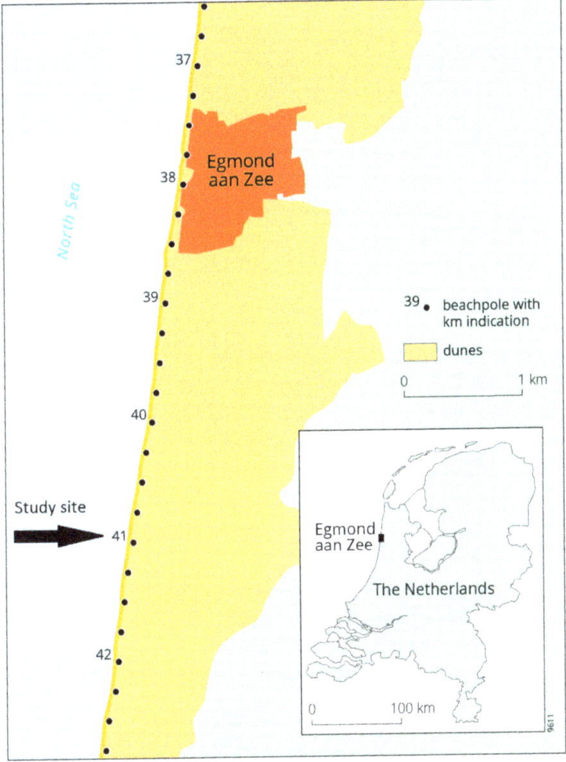

Figure 1. Location of study site. The number of a beach pole is the distance (in km) to the zero point at the northern end of the Dutch Holland coast.

The climate in the Netherlands is temperate humid with strong seasonal variability. The storm season is from October to February, with winds predominantly from Southwest to Northwest (i.e., onshore oblique). The monthly mean offshore significant wave height H_{m0} in the storm season is about 1.8 m, which is substantially higher than in the summer months (0.9 m) [41]. During northwesterly storms H_{m0} can increase to over 7 m. The tide is semi-diurnal with an average range of about 1.6 m but water levels vary with spring-neap cycles and wind conditions. During storm surges, which are most prominent during northwesterly storms, the beach may be flooded all the way to the foredune.

2.1.2. Aeolex-II

The data were collected in the framework of the Aeolex-II campaign, a field study on aeolian processes executed from 3 to 30 October 2017 by the Department of Physical Geography at Utrecht University. The overarching aim of the campaign was to collect data to investigate how properties of the wind, beach morphology, vegetation and sand (e.g., moisture content) control the timing and amount of aeolian sand transport toward and across the foredune. Measured variables comprised sea water level, groundwater elevation, ambient air temperature, in- and outgoing solar radiation, relative humidity, air pressure, rainfall, wind speed and direction, surface moisture content, beach and dune morphology, saltation intensity, aeolian mass flux and, finally, vegetation type and cover. Here a subset of the measured variables is used, as detailed in Section 2.1.3 below. Other parts of the Aeolex-II data were analyzed in, for example, De Winter et al. [19] and Schwarz et al. [45].

During Aeolex-II the offshore significant wave height H_{m0} peaked at 4 m on 5 October and almost 7 m on 29 October (Figure 2a), associated with gale-force (10-minute mean wind speeds of about 20 to 25 m/s) northwesterly to northern winds (Figure 2b,c). The first wind event took place during spring tide (Figure 2d) and caused minor dune erosion north of the study site, while the second event was during neap tidal conditions and no dune erosion was observed. Both events were associated with a storm surge of about 1 m (Figure 2e). From around 13 to 19 October fair weather conditions prevailed, with low wind speeds ($<\approx$10 m/s) and small significant wave heights (mostly < 1 m). All wave and water level data are from stations near IJmuiden, to the south of the study site, operated by the Dutch governmental organization Rijkswaterstaat; see Ruessink et al. [46] for precise locations. The wind data (at 10 m above ground) are from the meteorological station IJmuiden (WMO number 06225) operated by the Royal Netherlands Meteorological Institute (KNMI), located some 15 km to the south of the study site on a harbor mole [46].

Atmospheric variables that may influence surface moisture were measured during Aeolex-II with a meteorological station mounted to a beach pole on the dry beach at the field site. An overview of some meteorological variables collected during the field period is given in Figure 3. The air temperature (Figure 3a) was mostly between 10 and 15 °C. The exceptionally high temperatures on 15 and 16 October were caused by hurricane Ophelia, which hit Ireland during this period [47] but at the same time, resulted in very calm and warm conditions in the Netherlands. Figure 3b shows the relative humidity. The tipping bucket in the meteorological station on the beach malfunctioned during part of the campaign, and the rain intensity data in Figure 3c are from the meteorological station Wijk aan Zee (WMO number 06257) operated by the KNMI at N 52° 30′19.0″, E 04° 36′11.0″, about 9.5 km south of the study site. Finally, daily reference crop evaporation (Figure 3d), estimated by the KNMI with the Makkink equation [48] for the Wijk aan Zee station, was generally low: less than about 1.5 to 2 mm/day. The lowest values (<0.5 mm/day) correspond to overcast days. The actual evaporation from the beach was likely to be even less because of the strong limiting effect of unsaturated sand on evaporation [49,50].

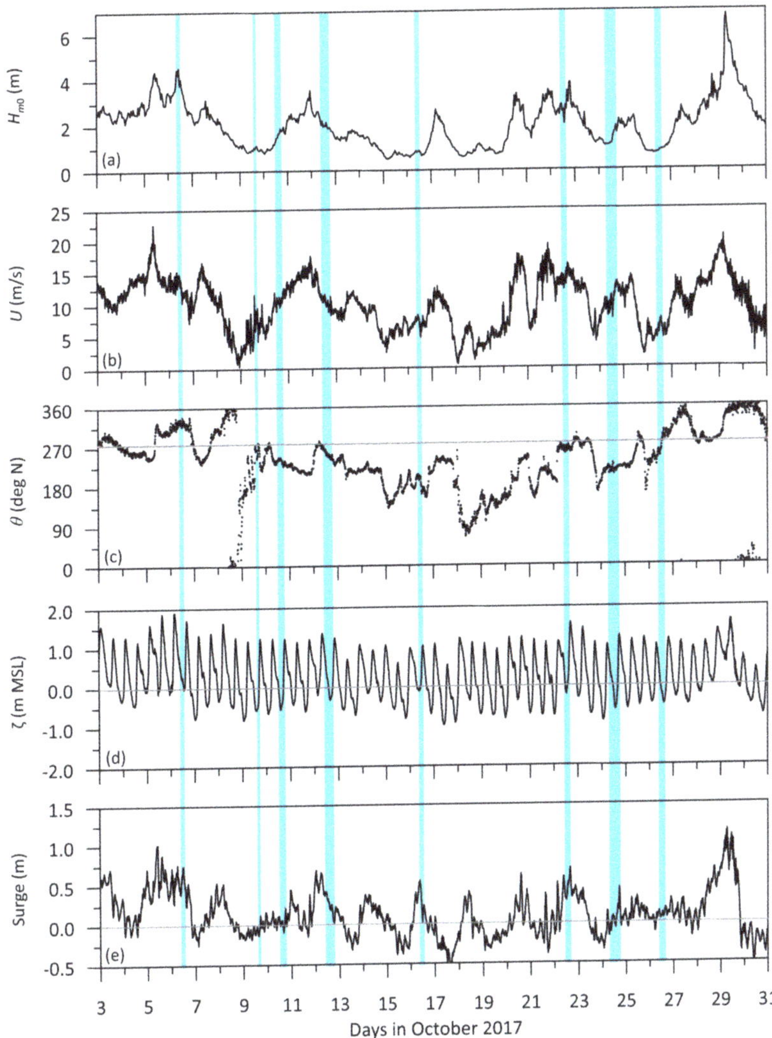

Figure 2. Time series of offshore (**a**) significant wave height H_{m0}, (**b**) wind speed U, (**c**) wind direction θ with respect to north (N = 0 and 360°, E = 90°, S = 180°, W = 270°), (**d**) water level ζ with respect to Mean Sea Level and (**e**) surge level during Aeolex-II. The sky-blue vertical bars represent the moments when surface moisture data were collected. The surge level was estimated as the difference between the measured and (predicted) astronomical water level. The horizontal gray line in (**c**) is the shore-normal direction, 277° N).

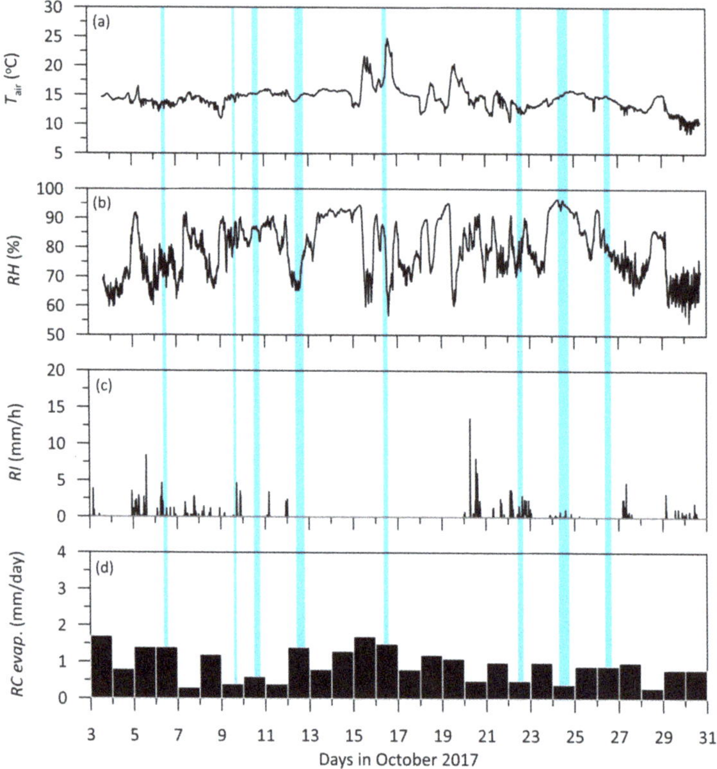

Figure 3. Time series of (**a**) air temperature T_{air}, (**b**) relative humidity RH, (**c**) rain intensity RI and (**d**) reference crop evaporation during Aeolex-II. The data in (**a**) and (**b**) were measured on the beach, those in (**c**) and (**d**) are from the nearby Wijk aan Zee meteorological station operated by the Dutch Royal Netherlands Meteorological Institute. The sky-blue vertical bars represent the moments when surface moisture data were collected.

2.1.3. Data Collection and Processing

Surface moisture measurements were conducted using a Delta-T theta probe (type HH2). The probe consists of a read-out unit in a waterproof housing that contains the electronics, which is connected to the sensor via an electric cable. The sensor has 4 stainless steel, 2-cm long rods that are inserted into the sediment. The original rod length is 6 cm but, following [29,51], the pins were shortened to measure the uppermost moisture content that is important to this study. The output of the probe ranges between 0 (dry) and 800 mV (fully saturated). To relate the probe's output to moisture content, a calibration was performed. Beach sediment samples at arbitrary locations on the beach were analyzed with the probe until the entire output range was captured. At each measuring location, a sample was collected with a 2-cm coring ring with the same diameter as the probe. The samples were bagged and immediately sealed for transport. Protocols of standard gravimetric moisture analysis were followed [52]; the wet mass of each sample m_w was determined after which the samples were dried in the oven for 24 h at 105 °C to achieve weight constancy. After drying, the mass of the samples was determined again, m_d. The gravimetric surface moisture content w_s [in %] of the sample then is $w_s = 100 \, (m_w - m_d) \, / \, m_d$. The calibration procedure was followed at the start and end of the of study period and found to give virtually identical results. The relationship between probe output and all w_s was described best with a fourth-order polynomial function, with a correlation-squared r^2 of 0.98 and a standard error of 1.17%. The curve clearly revealed that the device is most sensitive to surface moisture content in the range of 0 to ≈16%. Visual observations in the field showed that full saturation corresponds to $w_s \geq\approx$ 18%. During all measurements in both sampling strategies, the probe was inserted five times into the beach sediment, with the average probe output converted to the surface moisture content using the fourth-order calibration curve.

The moisture measurements were conducted on 6, 9, 10, 12, 16, 22, 24 and 26 October along a cross-shore array of groundwater wells (Figure 4). On 4 October eight wells, termed GW1 to GW8 from sea to land, were placed from the low-tide level to the top of the dry beach with a cross-shore spacing of 10 to 20 m. All sensors were set to store an instantaneous water level value with an interval of 5 min. The positioning of the wells implies that they measured sea water level ζ when the well was submerged by the tide and ground water level η otherwise. GW1–GW4 no longer stood perfectly upright by the end of the first storm. These 4 wells were removed on 7 October, with GW2–GW4 re-installed on 9 October at different cross-shore locations than before the storm. GW2 and GW4 did, however, not collect any data during their second deployment. The recorded time series were processed as detailed in [29,51], resulting in water level series with respect to Dutch ordnance level NAP (about equal to MSL). The wells were removed from the field on 28 October. In addition to moisture measurements at the well locations, measurements were also carried out at markers with an approximately 5-m spacing between the wells. Because these markers had to be removed with every incoming tide and were replaced the next measuring day, their spacing and number varied slightly each measurement day. The marker locations were obtained using an RTK-GPS system after each new transect was placed. Employing the sampling method as previously discussed with five measurements per marker, each location could be handled in about 45 s, implying that the entire cross-shore transect consisting of a maximum of 24 markers during low tide could be handled in less than 20 min. Depending on the inundation level of the beach the entire transect was measured in much shorter time.

A pressure transducer (PT) was deployed approximately 27 m seaward of GW2 (Figure 4). Its data, sampled at 5 Hz, were processed in 5-min. averaged values of sea water level (with respect to MSL), where each block of 5 min was centered around the sample moments of the GWs. Because of its position just landward of the low-tide shoreline, the PT was submerged during the entire tide in case of a storm surge, but during mid and high tide only otherwise. The PT location is used in this paper as the origin of a local cross-shore (x) coordinate system, with x positive in the landward direction (Figure 4).

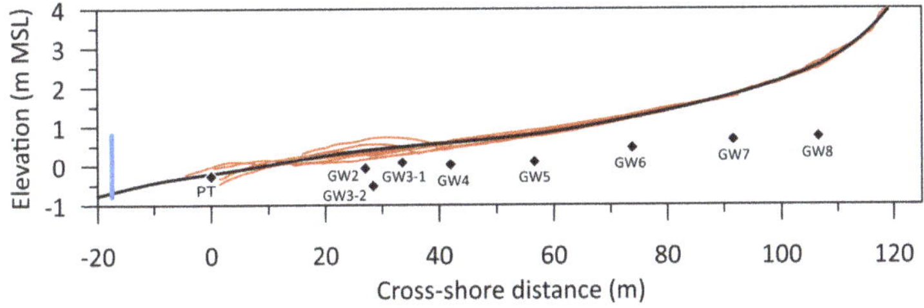

Figure 4. Bed elevation z versus cross-shore distance x on all moisture measurement days (six red-brown lines). The thick black line is the campaign-averaged cross-shore profile. Its extension seaward of $x \approx 0$ m is based on a jet-ski survey of the subtidal zone performed on 3 November. Also shown are the locations of the pressure transducer (PT) and the groundwater wells that operated during at least part of the campaign. GW3-1 and GW3-2 are GW3 prior to 7 October and after 9 October, respectively. See text for further explanation. The blue vertical line indicates the mean tidal range.

The cross-shore profile $z(x)$ along the GW-PT array, surveyed with an RTK-GPS, was typically featureless during and immediately after high-wave conditions and contained a bar-trough system in the intertidal zone otherwise. The campaign-average profile, based on 26 individual profiles (no surveys were performed on 18 and 29 October), was monotonically sloping in the landward direction (Figure 4), with an approximately 1:50 intertidal slope. Additional topographic data were collected before (23 September) and after (3 November) the campaign using an Unmanned Aerial Vehicle (UAV) equipped with a LiDAR system. The two flights covered an area that extended 1.4 km in the alongshore direction, roughly from the mid-tide shoreline to several tens of meter landward of the crest of the foredune. As detailed in Ruessink et al. [46], the collected 3D point clouds were processed into digital elevation models (DEMs) with a 1×1 m spatial resolution. The volume change of the seaward side of the foredune, taken from the 2.5 m MSL contour to the location of the change in slope [46], was positive (i.e., deposition) in the entire area. The volume gain at the PT-GW array, taken as the alongshore average of a 100-m wide section centered around the array, was about 3.3 $\mathrm{m^3/m}$ between 23 September and 3 November. This is a quite substantial number for an approximately one-month period, as, for example, the volume gain in the embryo dunes in 2017 as a whole (based on surveys in January and December 2017 [46]) was about 15 $\mathrm{m^3/m}$. The median grain size D at the array, determined by sieving sand samples collected at the end of the campaign, decreased in the landward direction from about 290 μm near the low-tide level to about 250 μm at the dune foot.

Finally, to aid in the interpretation of the GW data, we determined how each well was positioned with respect to the sea-land transition at any particular moment in time during Aeolex-II by calculating the total water level (TWL) on the beach. As in Cohn et al. [53], the TWL was taken as the sum of the offshore water level (tide and storm surge), the breaking-induced set-up and the 2% exceedance value of the swash. Here, the TWL was computed using the measured IJmuiden water level (Figure 2d), and the empirical Stockdon et al. [54] predictors for set-up and swash with the IJmuiden wave data (e.g., Figure 2a). As can be seen in Figure 5, the TWL peaked at 2.5 m +MSL during two successive high tides on 5 and 6 October, indicating that the swash may just have reached the most landward groundwater well (GW8) during the first storm. Following both peaks, the TWL dropped to the level of GW3 to 5 (≈0.5–0.7 m +MSL). For most of the campaign, however, GW7 and GW8 were above the high-tide TWL and all GWs were above the low-tide TWL. The maximum TWL (≈2.75 m +MSL) during Aeolex-II was reached on 29 October, but by this time all groundwater wells had, as aforementioned, already been removed from the beach.

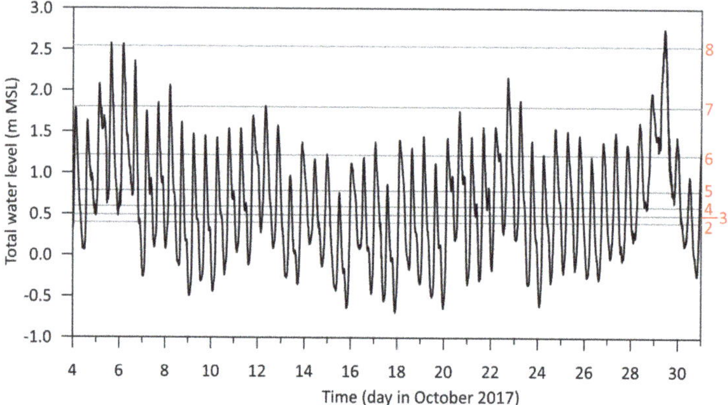

Figure 5. Total water level TWL versus time *t*. The horizontal lines are the bed level at the groundwater wells, indicated with the red numbers next to right vertical axis, based on the campaign-averaged cross-shore profile. GW3 is GW3-1. The line for GW3-2 is not shown separately as it is about equal to that for GW2, see Figure 4.

2.2. Model

2.2.1. Description

The model, called Aeolus, predicts the actual and potential cross-shore aeolian transport rate at the beach-dune transition. Aeolus comprises three modules: (1) a groundwater module to predict cross-shore and temporal water table fluctuations using the one-dimensional (cross-shore) non-linear Boussinesq equation extended with a run-up infiltration, (2) a moisture module that predicts surface moisture content from the predicted depth of the water table and a soil–water retention curve under the assumption of hydrostatic equilibrium, and (3) a fetch-based aeolian transport module in which the spatio-temporally varying surface moisture content is used as a supply-limiting factor. The groundwater and moisture module have been detailed in Brakenhoff et al. [29] and the full Aeolus model has been described in Hage et al. [39]. The Aeolus version used here differs from the version in [39] in that three modifications to the aeolian transport module were made: (1) the threshold for motion is computed explicitly rather than provided as a user-specified input parameter, (2) the scheme to relate surface moisture to the critical fetch has been simplified to equal that proposed in [18] and (3) the cosine-effect [12,13] has been included to facilitate the computation of foredune growth. In the following we provide the main equations of Aeolus.

The cross-shore and temporal water table fluctuations η' follow from

$$\frac{\partial \eta'(x,t)}{\partial t} = \frac{K}{n_e} \frac{\partial}{\partial x} \left\{ [d_a + \eta'(x,t)] \frac{\partial \eta'(x,t)}{\partial x} \right\} + \frac{U_l}{n_e}. \tag{1}$$

Here, the prime indicates a predicted value, t is time, d_a is the (constant) aquifer thickness, K is the hydraulic conductivity of the sand, n_e is the (non-dimensional) specific yield and U_l is the run-up infiltration flow rate per unit area [55]. The seaward boundary condition is a moving shoreline at location $x_{sh}(t)$ with elevation $\eta'_{sh}(x_s,t)$

$$\eta'_{sh}(x_{sh},t) = \zeta_0(t) + \zeta'_{sh}(t), \tag{2}$$

where ζ_0 is the offshore water level (Figure 2d) and $\zeta'_{sh}(t)$ is the breaking-induced set-up at the shoreline, for which the Stockdon et al. [54] parameterization is taken. The imposed landward boundary condition is $\partial \eta'/\partial x = 0$. The ground water module can handle monotonically increasing (in the landward direction) bed profiles only (i.e., no sandbar/trough profiles).

The depth of the water table beneath the bed surface, $h'(x,t) = \eta'(x,t) - z(x)$, is used to predict the surface moisture content $w'_s(x,t)$ using the Van Genuchten [56] soil–water retention curve,

$$w'_s(x,t) = w_{res} + \frac{w_{sat} - w_{res}}{\left[1 + (\alpha h'(x,t))^n\right]^{1-1/n}}. \tag{3}$$

Here, w_{res} and w_{sat} are the residual and saturated gravimetric water content, respectively, and α and n are free parameters. The use of Equation (3) implies an immediate response of surface moisture content to water table variations, without hysteresis effects [57]. Earlier work [29,51] has shown that this is a justified approach for the sediments at our field site.

Hage et al. [39] coupled the groundwater and surface moisture modules to a fetch-based aeolian sand transport model, thereby changing the conceptual fetch framework of Bauer and Davidson-Arnott [13] into a predictive model and generalizing the earlier predictive model of Delgado-Fernandez [18] that relied on a video-derived (i.e., observed) and cross-shore constant surface moisture content. Because w'_s decreases in the duneward direction from w_{sat} near the shoreline to w_{res} on the dry beach [29], Hage et al. [39] computed the downwind increase in the aeolian sand transport rate q (in kg m^{-1} s^{-1})

on each output time step of the moisture module with a spatially forward-stepping (i.e., from sea to land) equation,

$$q(i) = \begin{cases} \min\left[q_p,\; q(i_1 - 1) + q_p \sin\left(\frac{\pi}{2}\frac{F(i)}{F_c(i)}\right)\right] & \text{if } w_s(i) \leq w_{s,\max} \\ 0 & \text{otherwise,} \end{cases} \tag{4}$$

where q_p is the potential aeolian transport rate. For each grid point with w_s' less than a user-specified moisture threshold $w_{s,\max}$ (typically, 10% [18]) above which aeolian transport is inhibited, the critical fetch F_c is computed as

$$F_c = p(w_s') \times (4.38U - 8.23). \tag{5}$$

The rightmost term in this equation is the wind-speed U dependence of the critical fetch for dry sand as proposed by Delgado-Fernandez [18]. The term $p(w_s)$ describes the moisture dependent increase in the dry sand F_c, for which, as in Delgado-Fernandez [18], a step-function is adopted with $p = 1$ for $w_s' < 4\%$, $p = 1.25$ for $4\% \leq w_s' < 6\%$ and $p = 1.75$ for $6\% \leq w_s' \leq 10\%$. Please note that [39] used a slightly different $p(w_s')$, which demanded the rounding of w_s' to multiples of 0.5%; in the present scheme, this rounding is no longer necessary. In Equation (4) we denote a group of consecutive grid points with the same F_c by i; in general, there will be 3 F_c groups from sea to land, namely a group with $6\% \leq w_s' \leq 10\%$ ($i = 1$), with $4\% \leq w_s' < 6\%$ ($i = 2$) and with $w_s' < 4\%$ ($i = 3$). In each group, the fetch F is computed as the distance downwind of the last grid point seaward of the start of the group. This is indicated by the $i_1 - 1$ in Equation (4), with i_1 the first location index of the *i*th F_c group. The present F computation implies that F is reset to 0 each time F_c changes; that is, at $w_s' = 6$ and 4%. In case F exceeds F_c, $F = F_c$ is adopted. In essence, Equation (4) implies that the transport rate q in each group i thus equals the transport rate at its seaward boundary ($= q(i_1 - 1)$ in Equation (4)) increased with the fetch-based trend within the group ($= +q_p \sin((\pi/2)(F(i)/F_c(i)))$ in Equation (4)), up to a maximum of q_p.

The potential aeolian transport rate q_p (in kg m^{-1} s^{-1}) is predicted with Hsu [58] (see also Davidson-Arnott and Law [12]),

$$q_p = [-0.47 + 4.97 D_{\mathrm{mm}}] \left(\frac{U_*}{\sqrt{gD}}\right)^3 \times 10^{-5}. \tag{6}$$

The term $[-0.47 + 4.97 D_{\mathrm{mm}}]$ contains the dependence of q_p on the grain size D, which has to be specified in mm (hence, D_{mm}). The term between the large brackets is a Froude number, where $g = 9.81$ m s^{-2} is gravitational acceleration. Under the assumption of a logarithmic velocity profile, the shear velocity U_* is related to U as $U_* = aU$, where a depends on the measurement height of U and on the roughness length z_0. The Hsu [58] equation does not contain a threshold U_* for motion (U_{*t}). Following Delgado-Fernandez [18], Hage et al. [39] applied Equation (6) only when the wind speed exceeded a user-specified threshold value; otherwise, $q_p = 0$. Here, we estimated U_{*t} using Equation (7) and applied Equation (6) only when U_* exceeded U_{*t}. We adopted the Shao and Lu [59] formulation, modified following [60] to include the bed slope (β) effect on the initiation of motion,

$$U_{*t} = \Phi(\beta) A_N \sqrt{\frac{\rho_s - \rho_a}{\rho_a} gD + \frac{\gamma}{\rho_a D}}, \tag{7}$$

with

$$\Phi(\beta) = \sqrt{\cos\beta + \frac{\sin\beta}{\tan\Psi}} \tag{8}$$

and $\Psi \sim 33°$ the angle of repose. The non-dimensional parameter A_N is 0.111, $\rho_a = 1.25$ kg m^{-3} is the density of air, and γ is 2.9×10^{-4} N/m^{-1} [61].

The computation of q in Equation (4) is continued up to and including the beach-dune transition. The q at this transition is henceforth referred to as q_{bdt}. Its cross-shore (onshore) component, $q_{bdt,on}$, is (cosine-effect)

$$q_{bdt,on} = q_{bdt} \cos \theta_{SN},$$

(9)

in which θ_{SN} is the wind direction with respect to the shore-normal direction. The volume growth of the foredune ΔV during a given time interval can be computed as

$$\Delta V = \frac{\sum (q_{bdt,on}(t)\Delta t)}{(1-p)\rho_s}.$$

(10)

Here, Δt is the model output time step, p is bed porosity (typically, 0.4) and ρ_s is the sediment density (for quartz sand, $\rho_s = 2650$ kg m^{-3}). The computation of ΔV implies that all sand that crosses the beach-dune transition is deposited on the foredune. This is a realistic assumption for the present study site because of the dense vegetation on the upper part of the foredune [45,46,62,63].

2.2.2. Set-Up

The settings of the groundwater module were taken from Hage et al. [39] ($K = 4.63 \times 10^{-4}$ m/s, $n_e = 0.3$ and $d_a = 15$ m), which, in turn, were based on earlier modeling work for the study site (e.g., [64]). The seaward boundary condition was taken as the sum of the offshore water level shown in Figure 2d and the set-up computed with the Stockdon et al. [54] predictor using the offshore wave data collected at IJmuiden (see Figure 2a for the wave height). The campaign-averaged cross-shore profile (Figure 4) was used as the (time-invariant) cross-shore profile $z(x)$, with a cross-shore grid spacing of 0.5 m. The time step in the groundwater module was set to 2 s, with cross-shore profiles of water table fluctuations $\eta'(x)$ written as output every 10 min (i.e., $\Delta t = 600$ s). The settings for the surface moisture module (w_{res}, w_{sat}, α and n) will be based on the fitting the soil–water retention curve, Equation (3), to the Aeolex-II observations as presented in Section 3.1 below. The model was run from 1 September to 4 November 2017, and thus encompasses the entire period between the two UAV-LiDAR surveys. The approximate 3-week period prior to 23 September served as model spin-up time.

The fetch-based aeolian transport module was forced with the wind data measured at IJmuiden (Figure 2b,c), with onshore winds modified to local (i.e., on the beach) conditions to allow for the effect of the high and steep foredune on the wind field. De Winter et al. [19], a study largely based on the Aeolex-II wind data, illustrated that the wind speed on Egmond beach corrected to the same height as at IJmuiden (10 m) was substantially lower than that measured at IJmuiden. The ratio of the local to regional (IJmuiden) wind speed was lowest ($\approx$0.7) for onshore winds to increase gradually to near 1 for alongshore winds (Figure 6a). De Winter et al. [19] further indicated that the wind direction on the beach was essentially equal to that measured at IJmuiden. In contrast, the wind at the beach-dune transition was often more alongshore directed (Figure 6b), with the deflection angle being largest $\pm(10$–$15^\circ)$ when the regional wind approached the beach at about $\pm45^\circ$. Therefore, in the computation of the fetch (i.e., in Equation (4)) we used the regional wind direction, but the alongshore deflected direction in the computation of the onshore component of the aeolian transport rate at the beach-dune transition (i.e., in Equation (9)). When the IJmuiden wind was blowing offshore, q_p was set to 0. This implies that we ignore the contribution of offshore winds to foredune accretion due to wind flow reversal on the seaward side of the foredune [20,24]; extensive video observations of Egmond beach did not reveal any noteworthy aeolian activity under offshore winds [65], hence our choice to restrict the Aeolus model to onshore wind conditions. The grain size D was taken as 250 µm, and the moisture threshold $w_{s,max}$ as 10%. Based on an analysis of TWL values for Egmond between January 2015 and 2019 [46], $z_{bdt} = 2.5$ m +MSL was adopted. Finally, extensive observations during Aeolex-II with a vertical array of 5 cup anemometers located just above the high-tide level revealed a typical roughness length z_0 of about 1×10^{-4} m. Therefore, the ratio a between the wind shear velocity U_* and wind speed U at $z = 10$ m above ground, was set to $\kappa \log(z/z_0) = 0.0356$, using 0.41 for the Von Karman constant κ. This a is about 11% lower than the commonly applied $a = 0.04$ proposed by Hsu [66].

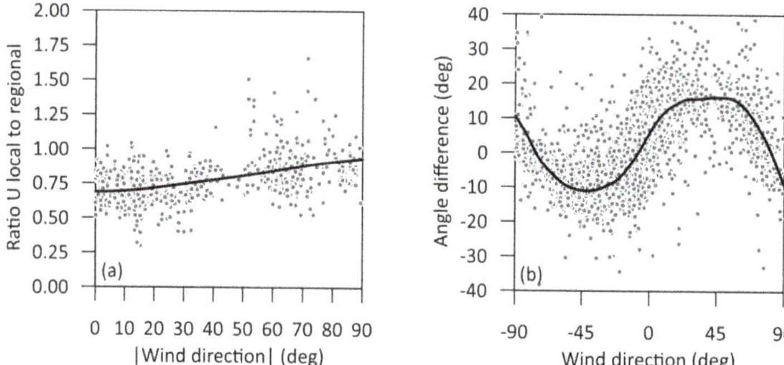

Figure 6. Correction of regional (IJmuiden) to local (Egmond beach) wind data. (**a**) shows the ratio of local to regional wind speed versus the absolute value of the regional wind direction with respect to the shore normal, $|\theta_{SN}|$. Local wind speeds were measured at 0.9 m above the bed and were corrected to the same height above ground (10 m) as where the regional values were measured, assuming a logarithmic velocity profile and a roughness length z_0 of 1×10^{-4} m. The panel contains all 730 observations (dots) taken on the intertidal beach during Aeolex-II and the smooth trend line. (**b**) shows the observed (dots; Aeolex-II) difference between the wind direction at the beach-dune transition and at IJmuiden versus the regional wind direction with respect to the shore normal, θ_{SN}. The line is the smooth trend line. This panel contains about 2950 observations. Shore-normal onshore wind (i.e., coming from 277 °N) has $\theta_{SN} = 0°$; negative and positive angles on the x-axis in (**b**) represent winds coming the southwest and northwest, respectively. Because the local to regional wind speed ratio (**a**) was symmetric with respect to θ_{SN}, the absolute value of the direction was used to better constrain the trend line. The angle steering in (**b**) was not perfectly symmetric, perhaps because of the embryo dunes to the south of the instrument array [45]. See [19] for details of data collection and processing. We note that the results in (**a**) are different from those in [19] in the sense that they provided correction values in 45° bins.

3. Results

3.1. Observations

We first focus our attention on the measurements on 6 October (Figure 7), which were performed after the peak in the storm surge on the previous day and night (Figure 2e). The PT recorded a high-tide water level of 1.9 m +MSL around 03:45 (Figure 7c). The estimated TWL of 2.5 m (Figure 5) indicates that as aforementioned, the largest swashes may just have reached GW8 during this high tide. The groundwater at GW8 peaked near 2.2 m +MSL (Figure 7c), which was about 0.7 m higher than recorded on 4 October before the storm. GW7 was submerged during this high tide (the water level was above the bed elevation at GW7; Figure 7c). After GW7 had emerged near 08:00, the ground water level gradually fell to about 1.5 m +MSL, i.e., to about 0.25 m beneath the bed surface. The water level at GW6 remained constant with time at about 1.2 m +MSL after emergence; this water level equals the bed elevation at GW6. Similar behavior is also seen at the GWs lower on the beach. These constant water level values indicate the presence of a seepage face, which was also visually observed during the measurements as a glassy beach surface. A slight drop in water level at GW6 around 12:00 probably indicates that the point where the water table intersected the beach face moved seaward past GW6 around this time. Because the lower GWs were temporarily removed from the field, we cannot say for how many tides the water table and the (sea) tide remained decoupled after the surge of 6 October. The water levels at GW3-2 remained constant at beach surface elevation (when not submerged by the tide) between 9 and 12 October (not shown), after which they reduced until the submergence by the next tide (e.g., see Figure 8c below). The moisture measurements were performed where the beach was not completely saturated, from the dune foot to approximately the location of GW7. As can be seen in Figure 7a, w_s was low (<4%) and constant with time landward of GW8 (i.e., landward of the largest swashes during high tide). Between GW7 and GW8 w_s dropped by approximately 10% during the 4-h measurement period, widening the dry beach by approximately 5 to 10 m to $x = 98$ m. In more detail, the part of the beach with the largest reduction in w_s moved seaward during falling tide, from near $x = 105$ m between 08:30 and 10:15, to $x = 95$ m between 10:15 and 12:42. The remainder of the beach, seaward of GW7 ($z < 1.7$ m), remained fully saturated during the entire measurement period.

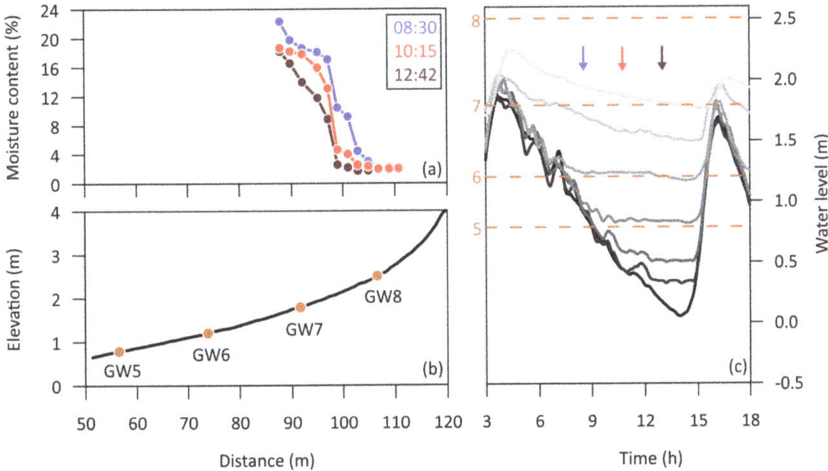

Figure 7. (**a**) Surface moisture content w_s at three time intervals during falling tide and (**b**) bed elevation z versus cross-shore distance x on 6 October, at the end of the storm surge that peaked on the previous and night. (**c**) shows time series of water level at six groundwater wells, GW8 (light gray) down to GW3 (dark gray). The black line is the water level at the PT. The three vertical arrows mark the moment of moisture sampling, with the same color coding as in (**a**). The four dashed brown lines are the bed elevation at (top to bottom) GW8 to GW5, see also (**b**).

Figure 8 illustrates the surface moisture and water level measurements during a falling tide without a surge (26 October). The maximum water level recorded by the PT during the preceding high tide was just below 1 m +MSL (Figure 8c) and the TWL estimated for this moment (Figure 5) suggests that the upper limit of the swash zone was near GW6. The water levels at GW8 and GW7 were substantially lower than on 6 October and barely showed any variation with the tide. The surface moisture content at GW7 and GW8 was below 4% and did not vary with time (Figure 8a). At the start of the measurements, just after high tide, w_s rose sharply to about 20% within 15–20 m between GW7 and GW6. With time, the intertidal beach dried, with the lowest w_s around the high-tide level (GW5 to GW6) and on the crest of the intertidal bar near $x = 20$ m and with w_s remaining largest in the trough near $x \approx 30$ m (Figure 8a). The reduction in w_s coincided with the falling of the ground water level (i.e., increase in ground water depth) with time of the wells in the intertidal zone (GW3-2, GW5 and GW6). Despite the overall drying, w_s remained above 10% and at many places above 15%. This suggests that even under these conditions most of the intertidal beach remained too wet to sustain aeolian transport.

The w_s observations of all days without a surge confirm the observations of 26 October and further illustrate that under such conditions the beach can broadly be divided into three cross-shore moisture zones (Figure 9a), consistent with earlier studies conducted under low-wave conditions [29,51,67–70]. These zones were here $x <\approx 10$ m (lower beach), $x \approx 10$–80 m (mid-beach) and $x >\approx 80$ m (upper beach). The lower beach, corresponding to the intertidal beach with $z < 0$ m MSL, was always saturated. The mid-beach, which is the intertidal beach with $z > 0$ m MSL, experienced saturated conditions immediately after emergence during the falling tide. With time, the surface moisture content dropped to about 10%, or to 5% in the presence of a sandbar (here, between $x \approx 15$ and 40 m). The upper beach had mostly dry (0–5%) conditions, especially at the higher parts of the upper beach toward the dune foot. This cross-shore moisture zonation was disturbed severely by the presence of the surge on 6 October. Then, as described in connection with Figure 7, dry conditions were only found in a narrow strip just above the maximum TWL, while most of the remaining (sub-aerial) beach remained saturated because of exfiltrating groundwater. Moisture values dropped to more intermediate values in a 10–15 m wide zone beneath the maximum TWL only.

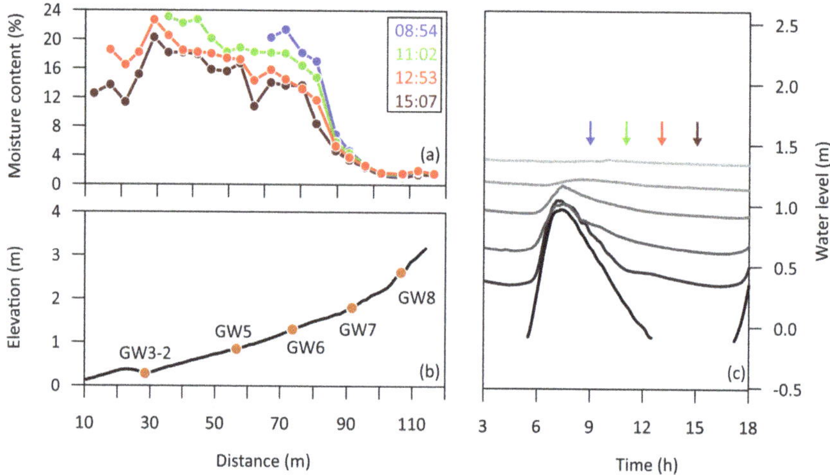

Figure 8. (**a**) Surface moisture content w_s at four time intervals during falling tide and (**b**) bed elevation z versus cross-shore distance x on 26 October, a day without a surge. (**c**) shows time series of water level at five groundwater wells, from light to dark gray in seaward direction: GW8, GW7, GW6, GW5 and GW3-2; for their precise location, see (**b**). The black line in (**c**) is the water level at the PT; no data are available at low tide because the PT was not submerged. The four vertical arrows mark the moment of moisture sampling, with the same color coding as in (**a**).

The observed spatio-temporal variability in surface moisture content (Figures 7a, 8a and 9) was, consistent with earlier studies [29,51,57,70,71], controlled largely by the depth h of the groundwater beneath the beach surface. This is illustrated in Figure 10 based on all moisture measurements collected next to a well. When h was less than 0.15 to 0.2 m, the sand remained saturated ($w_s \approx 18$–22%), implying that the capillary fringe still intersected the beach surface. With increasing h, w_s rapidly decreased until $h \approx 0.7$ m where the sand reached the field capacity ($w_s \approx 1.5$%). From this depth, surface moisture content was no longer affected by groundwater depth. The often-used moisture threshold of $w_s \approx 10$% above which aeolian transport ceases, corresponds here to $h \approx 0.3$ m. The data from two measurement days are highlighted in Figure 10 to examine potential atmospheric effects. The first day (red dots) is 16 October, when evaporation was expected to be high because of the high air temperature and sunny conditions. However, the w_s values for this day fall within the scatter of the other days, even for large h, suggesting that evaporation did not affect moisture content notably. The second day (blue dots) is 22 October, when the measurements were carried out during a light rain. Again, the moisture observations showed the same dependence on groundwater depth as on the other days, except for the observations with $h > 1$ m on the upper beach. Here, w_s was a few percent larger than on the other days. Overall, our results imply that in our data groundwater processes were more important in modulating the surface moisture content than atmospheric processes. The relationship between h and w_s can be described well using Equation (3), see Figure 10. A non-linear fit with

$w_{res} = 1.5\%$ and $w_{sat} = 18.2\%$ resulted in $\alpha = 3.18$ m^{-1} and $n = 5.47$. This best-fit curve, used in the Aeolus modeling, has a mean square error of about 2.9%.

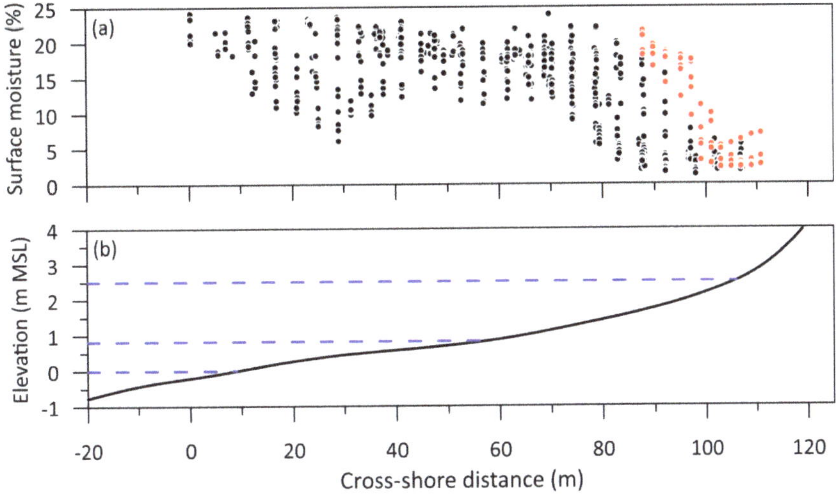

Figure 9. (**a**) Surface moisture content w_s versus cross-shore distance x based on all measurement days (848 observations). The observations made on 6 October (immediately after the first storm) are shown in red. (**b**) shows the campaign-averaged cross-shore profile. The blue dashed lines are (from bottom to top) mean sea level, the normal mean high-tide level and the maximum total water level on 6 October.

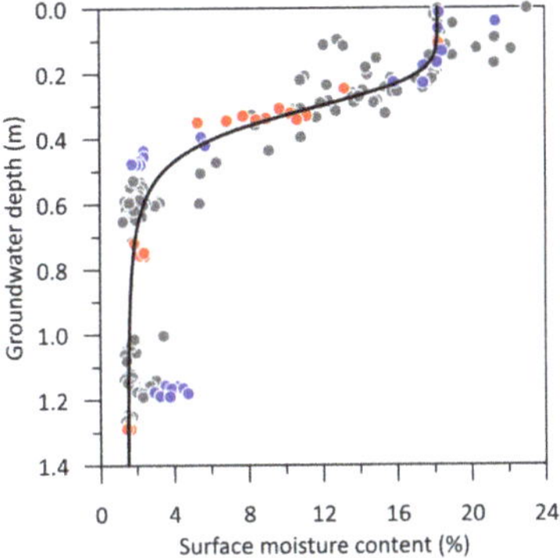

Figure 10. Groundwater depth h versus surface moisture content w_s. The red dots are the measurements of 16 October, when the air temperature was rather high because of hurricane Ophelia (Figure 3a). The blue dots were obtained on 22 October in a light rain (Figure 3c). The line is the best-fit water retention curve, Equation (3). The total number of observations is 194.

3.2. Modeling

Consistent with the observations in Figure 7, the groundwater-moisture model predicted most of the beach to remain saturated after the storm-surge high tide on 6 October (Figure 11; $x < \approx 80$ m, $w'_s = w_{sat}$). Surface drying was predicted to remain restricted to an approximately 20 m wide zone ($x \approx 80$–100 m). This is slightly wider than in the observations, possibly because of the neglect of groundwater recharge from the dune in the landward boundary condition of the groundwater module [64]. Consequently, the predicted groundwater in the upper beach sat slightly deeper beneath the bed surface and fell faster with time than in the observations. This resulted in a zone with reducing surface moisture content that extended further seaward from the fully dry zone with time than in the observations. The predicted deeper groundwater also resulted in a further seaward location of the groundwater exit point on the beach, between GW6 and GW5 as opposed to between GW7 and GW6 in the observations. However, in the predictions the groundwater depth at GW6 remained less than 0.2 m and, accordingly, w'_s still equaled w_{sat} (Figure 11). Despite these groundwater-induced discrepancies, the overall spatio-temporal change in surface moisture after the surge is predicted fairly well.

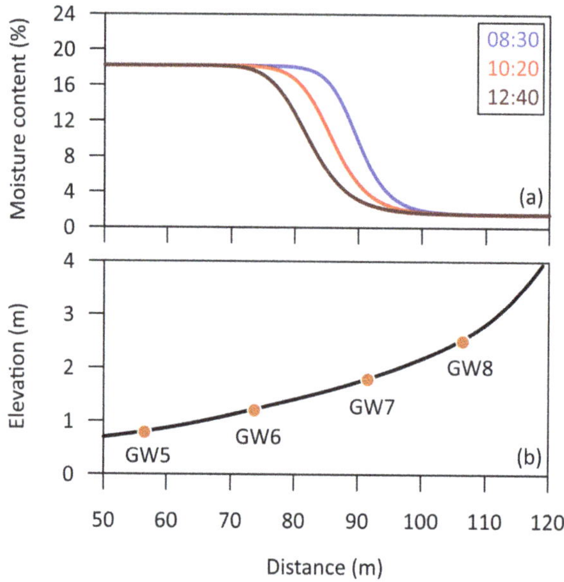

Figure 11. (**a**) Predicted surface moisture content w'_s at the three observational time intervals during falling tide on 6 October versus cross-shore distance x, compare to Figure 7a. (**b**) contains the campaign-averaged bed elevation z versus cross-shore distance x with the location of the upper 4 GWs. Predictions in (**a**) resulted from predicted groundwater elevations (Equation (1)) and the soil–water retention curve (Equation (3)) with $w_{res} = 1.5\%$, $w_{sat} = 18.2\%$, $\alpha = 3.18$ m^{-1} and $n = 5.47$ (Figure 10).

During the storm on 5 and 6 October the onshore potential transport $q_{p,on}$ followed the wind speed (compare Figure 12a and Figure 12e) and peaked near 0.03 kg/m/s. In contrast, the actual transport $q_{bdt,on}$ varied with the tide and was often considerably lower than $q_{p,on}$ (Figure 12a). During the two highest tides, aeolian transport was predicted to nearly cease ($q_{bdt,on}/q_{p,on} < 0.2$; Figure 12b). In more detail, the ratio $q_{bdt,on}/q_{p,on}$ was lowest just after each high tide. The temporal delay of the high-tide water table in the upper beach with respect to the sea-surface tide caused a delay in wetting of the sand, and hence $w_{s,max}$ was at its most landward position shortly after high tide. The predicted foredune growth ΔV based on $q_{bdt,on}$ (with $p = 0.4$ and $\rho_s = 2650$ kg/m^3) for 5 and 6 October was almost 0.49 m^3/m, about 66% of the potential foredune growth of 0.74 m^3/m. This substantial reduction highlights the relevance of the storm surge in limiting onshore aeolian sand transport and, hence, foredune growth. On the day before (4 October) and after (7 October) the storm surge the onshore transport at the beach-dune transition was predicted to be virtually equal to the potential onshore transport. Under these conditions the supply was thus primarily controlled by wind characteristics rather than by supply-limiting factors. Only around high tide the transport was slightly supply limited, with $q_{bdt,on}/q_{p,on}$ reducing to 0.7–0.8 (Figure 12b). This is consistent with video observations of aeolian activity at Egmond beach [36], from which a shift from unlimited to supply limited conditions from high to low tide was sometimes inferred. For the period between the two UAV-LiDAR surveys, Aeolus predicted an actual and potential ΔV of about 4.1 and 4.6 m^3/m, respectively. The predicted actual growth is rather close to (albeit somewhat above) the observed $\Delta V = 3.3$ m^3/m in the 100-m alongshore section centered around the instrument array. Possibly, the over-prediction is due to the wider cross-shore zone with drying sand after the surge than in the observations, or to the neglect of other supply-limiting factors such as rainfall.

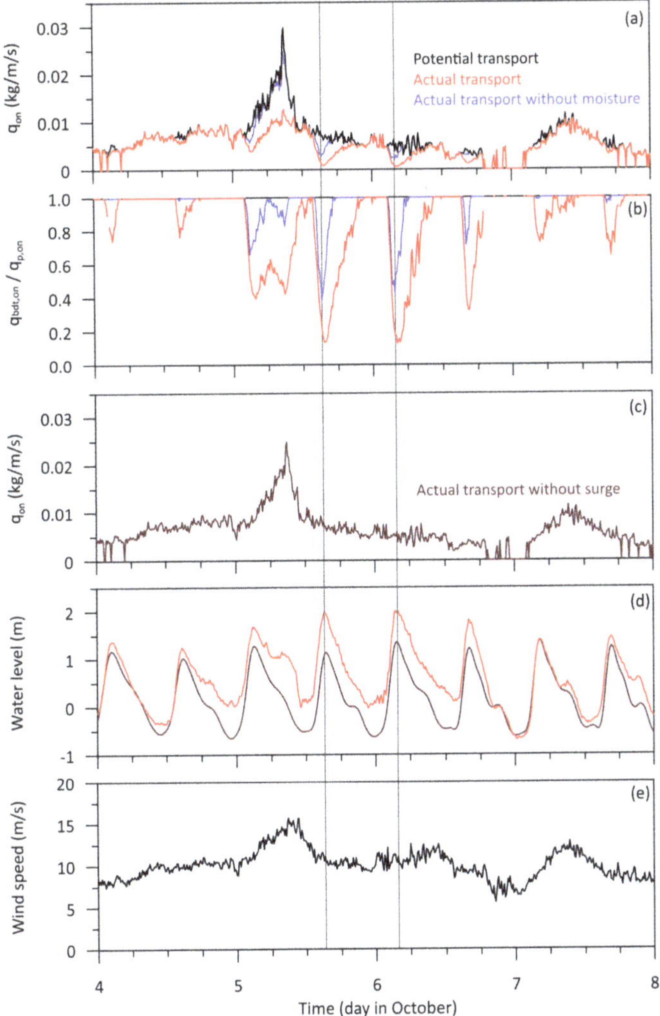

Figure 12. Results of aeolian transport modeling for a selected 4-day period in early October 2017. (**a**) shows the onshore potential transport rate $q_{p,\mathrm{on}}$ (black line) and the onshore actual transport rate at the beach-dune transition $q_{\mathrm{bdt,on}}$ (red line). The blue line in (**a**) is $q_{\mathrm{bdt,on}}$ in case the effect of surface moisture on q is ignored (i.e., $w_s = 0$ at all non-submerged locations). (**b**) illustrates the ratios between $q_{\mathrm{bdt,on}}$ and $q_{p,\mathrm{on}}$ with (red line) and without (blue line) surface moisture. (**c**) shows the results of another scenario (dark brown line) in which the astronomical tide is used as seaward boundary condition in the groundwater module (i.e., no storm surge and run-up; surface moisture is included). The applied astronomical boundary condition in (**c**) is the dark brown line in (**d**), with the red line in (**d**) being the seaward boundary condition applied in both model simulations shown in panel (**a**). The wind speed on the beach is given in (**e**). The two vertical lines through all panels are the two moments with the highest tides on 5 and 6 October.

4. Discussion

The predicted $q_{bdt,on}/q_{p,on} < 1$ during the surge may have arisen in two complimentary ways. First, the surge reduced beach-width and, consequently, the maximum available fetch. Secondly, the surge caused most of the beach that emerged during falling tide to remain saturated and to not become available for aeolian transport. To disentangle these two effects, we ran two additional simulations, the results of which are also shown in Figure 12. In the first simulation we took the groundwater predictions of the default run (thus including the surge) but we set the moisture content of the emerged beach to 0%. In this way, aeolian transport can commence immediately at the shoreline and the critical fetch is determined by wind speed alone. This run thus shows the effect of the surge-induced surface moisture effects given the surge-induced reduction in beach-width. As can be seen in Figure 12a (blue line), the onshore aeolian transport at the beach-dune transition is now predicted to be much closer to the onshore potential transport. The ratio $q_{bdt,on}/q_{p,on}$ reaches less low values at two high tides on 5 and 6 October ($\approx$0.4 compared to 0.1), is less than 1 for a reduced time interval around high water only, and is no longer delayed with respect to the high tide (Figure 12b). The ΔV for 5 and 6 October in this simulation was about 0.68 m^3/m, 0.19 m^3/m larger than in the default run and now 92.5% of the potential ΔV. Further note that transport limitations at high tide were predicted to largely vanish on the other days shown in Figure 12. In the second additional simulation, the groundwater module was run with the astronomical tide as seaward boundary condition (Figure 12d), i.e., without the surge and without waves, but surface moisture was computed for the entire emerged beach. This simulation thus illustrates the effect of the surge-induced reduction in beach-width. We found that without the surge the maximum fetch was nearly always sufficient for $q_{bdt,on}$ to reach $q_{p,on}$ at the beach-dune transition (compare Figure 12c to Figure 12a), with a resulting ΔV for 5 and 6 October that was only 2.4% lower than the potential ΔV. On the whole, these simulations illustrate that the influence of the storm surge in limiting aeolian transport at the beach-dune transition is due to fetch limitations imposed by the reduced beach-width and the saturation of the emerged (intertidal) beach. The low slope (1:50) and low conductivity ($K = 4.63 \times 10^{-4}$ m/s) associated with the fine-medium grain size ($D = 250$ μm) at our site may have favored the large seepage face extent [72] and hence its importance for supply limitation. Future work is necessary to establish the role of seepage face formation and associated saturated moisture conditions in limiting aeolian supply on steeper and coarser-grained beaches.

The model predicted that aeolian supply was unlimited ($q_{bdt,on}/q_{p,on} = 1$) under non-surge conditions, except for a restricted amount of time around several high tides. Indeed, the vast majority of the difference between the predicted actual (4.1 m^3/m) and potential (4.6 m^3/m) deposition over the Aeolex-II study period was due to the storm surges on 5–6 and 29 October. Our model results thus corroborate earlier field studies [11,34,35] that inferred storm surges to be the primary condition under which most of the over-prediction of foredune growth with a potential transport equation arises. Future work is necessary to see whether the inferences we made based on the approximately 1.5-month long study period also apply to seasonal and annual time scales. Model predictions on these larger time scales may reveal how supply limitation depends on surge characteristics, such as surge duration, magnitude and timing with respect to high tide. It is further possible that certain conditions that are not in our data set, such as strong onshore winds without substantially elevated water levels, may also result in fetch and hence supply limitations at our site [36].

Although the reasonable agreement between measured (3.3 m^3/m) and predicted (4.1 m^3/m) ΔV is encouraging, also for longer term predictions, there is certainly scope for model improvement. First, the landward boundary condition for the groundwater module could be improved to handle recharge from the foredune. Also, the groundwater module can handle monotonically increasing bed profiles only and, accordingly, the effect of an intertidal sandbar on surface moisture content (Figure 9; [73]) and on aeolian sand transport [74] cannot be predicted. Secondly, the surface moisture module is driven entirely by groundwater fluctuations; meteorological effects, such as evaporation and rainfall, are ignored. For the present data, we found no evidence that evaporation was relevant (Figure 10), but the range of conditions encountered was admittedly small. It is also possible that we missed evaporation effects because the applied Delta-T theta probe provided moisture content averaged over the top 2 cm of beach sand. The possible drying of the uppermost layer of grains on especially sunny and windy days was thus not measured. The role of rainfall on aeolian transport is unclear. For example, in the transport model of Duarte-Campos et al. [75] aeolian transport is set to 0 when rain intensity exceeds 0.05 m/h, while field measurements reveal non-zero aeolian transport under larger rain intensities [20,76,77]. Also, during Aeolex-II we visually observed substantial aeolian transport during such moderate rainfall, especially when winds were strong. In addition, van Dijk et al. [78] illustrated that severe rain (>2 mm/h) may initiate aeolian transport by splash, although this effect is probably limited to the very early stages of a shower [76]. Thirdly, the Aeolus model assumes a homogeneous sediment. Sand supply limitations often observed under heterogeneous sediments [31,32], typical of nourished beaches and likely caused by the development of a lag deposit, are not included. Fourthly, wind speeds may reduce across the beach [19,25], especially when the wind is cross-shore directed. This variation, which is not accounted for in Aeolus, could influence aeolian transport rates at the beach-dune transition. Finally, it would be useful to include other equations to compute q_p [79,80] to examine the sensitivity of ΔV predictions to the potential transport equation.

5. Conclusions

Our observations at the low-gradient (1:50) beach of Egmond aan Zee, Netherlands, illustrate that spatio-temporal surface moisture content immediately after a surge is controlled by the maximum total water level during the surge peak and the strongly elevated groundwater levels in the upper beach. The latter results in the decoupling of the groundwater from the sea water level toward low tide and hence the development of a seepage face. The surface moisture content is time-invariant and low (<4%; dry sand) above the maximum total water level, while the sand seaward of the groundwater exit point remains saturated with time ($\approx$18%). Drying of beach sand is restricted to a narrow (here, 10–15 m) cross-shore zone between the maximum water level and the groundwater exit point. The beach-width reduction by the surge-induced inundation and the persistent saturation of the emerging beach are both predicted to contribute to fetch limitations that substantially reduce aeolian supply to the foredune, for the 2-day surge period examined here to about 66% of the potential supply. Because during non-surge conditions the supply is almost always equal to the potential supply, our model results provide quantitative support for the common assertion that storm surges are the primary condition that cause the overestimation of measured foredune growth with a potential transport equation. Future work is needed to test the generality of our findings for other beach slopes and longer prediction intervals.

Author Contributions: Conceptualization, J.T.T., J.J.A.D. and G.R.; methodology, J.T.T. and G.R.; field investigation, J.T.T.; writing—original draft preparation, J.T.T. and G.R.; writing—review and editing, all authors; visualization, J.T.T. and G.R.; supervision, J.J.A.D., C.S.S. and G.R.; project administration, G.R.; funding acquisition, G.R. All authors have read and agreed to the published version of the manuscript.

Funding: This work is part of the Vici research programme "*Aeolus* meets *Poseidon*: wind-blown sand transport on wave-dominated beaches" with project number 13709, which is financed by the Dutch Research Council (NWO).

Acknowledgments: We thank Bas van Dam, Marcel van Maarseveen, Henk Markies, Mark Eikelboom and Arjan van Eijk for excellent technical support during Aeolex-II; and Corinne Böhm, Job van Beem and Jorn Bosma for their indispensable help during the measurements and simply a great time in the field.

Conflicts of Interest: The authors declare no conflict of interest. The funders had no role in the design of the study; in the collection, analyses, or interpretation of data; in the writing of the manuscript, or in the decision to publish the results.

References

1. Martínez, M.L.; Psuty, N.P.; Lubke, R.A. A Perspective on coastal dunes. In *Coastal Dunes, Ecology and Conservation*; Martínez, M.L., Psuty, N.P., Eds.; Springer: Berlin/Heidelberg, Germany, 2004; Volume Ecological Studies, 171, Chapter 1.

2. Hesp, P.A. Foredunes and blowouts: Initiation, geomorphology and dynamics. *Geomorphology* **2002**, *48*, 245–268. [CrossRef]

3. Carter, R.W.G.; Stone, G.W. Mechanisms associated with the erosion of sand dune cliffs, Magilligan, Northern Ireland. *Earth Surf. Process. Landf.* **1989**, *14*, 1–10. [CrossRef]

4. Castelle, B.; Marieu, V.; Bujan, S.; Splinter, K.D.; Robinet, A.; Sénéchal, N.; Ferreira, S. Impact of the winter 2013–2014 series of severe Western Europe storms on a double-barred sandy coast: Beach and dune erosion and megacusp embayments. *Geomorphology* **2015**, *2015*, 135–148. [CrossRef]

5. De Winter, R.C.; Gongriep, F.; Ruessink, B.G. Observations and modeling of alongshore variability in dune erosion at Egmond aan Zee, the Netherlands. *Coast. Eng.* **2015**, *99*, 167–175. [CrossRef]

6. Van Thiel de Vries, J.S.M.; van Gent, M.R.A.; Walstra, D.J.R.; Reniers, A.J.H.M. Analysis of dune erosion processes in large-scale flume experiments. *Coast. Eng.* **2008**, *55*, 1028–1040. [CrossRef]

7. Thornton, E.B.; MacMahan, J.; Sallenger, A.H. Rip currents, mega-cusps, and eroding dunes. *Mar. Geol.* **2007**, *240*, 151–167. [CrossRef]

8. Roelvink, D.; Reniers, A.; van Dongeren, A.; van Thiel de Vries, J.; McCall, R.; Lescinski, J. Modelling storm impacts on beaches, dunes and barrier islands. *Coast. Eng.* **2009**, *56*, 1133–1152. [CrossRef]

9. Splinter, K.D.; Palmsten, M.L. Modeling dune response to an East Coast Low. *Mar. Geol.* **2012**, *329–331*, 46–57. [CrossRef]

10. Castelle, B.; Bujan, S.; Ferreira, S.; Dodet, G. Foredune morphological changes and beach recovery from the extreme 2013/2014 winter at a high-energy sandy Coast. *Mar. Geol.* **2017**, *385*, 41–55. [CrossRef]

11. Delgado-Fernandez, I.; Davidson-Arnott, R. Meso-scale aeolian sediment input to coastal dunes: The nature of aeolian transport events. *Geomorphology* **2011**, *126*, 217–232. [CrossRef]

12. Davidson-Arnott, R.G.D.; Law, M.N. Measurements and prediction of long-term sediment supply to coastal foredunes. *J. Coast. Res.* **1996**, *12*, 654–663.

13. Bauer, B.O.; Davidson-Arnott, R.G.D. A general framework for modeling sediment supply to coastal dunes including wind angle, beach geometry, and fetch effects. *Geomorphology* **2002**, *49*, 89–108. [CrossRef]

14. Keijsers, J.G.S.; Poortinga, A.; Riksen, M.J.P.M.; Maroulis, J. Spatio-temporal variability in accretion and erosion of coastal foredunes in the Netherlands: Regional climate and local topography. *PLoS ONE* **2014**, *9*, e91115. [CrossRef] [PubMed]

15. Strypsteen, G.; Houthuys, R.; Rauwoens, P. Dune volume changes at decadal timescales and its relation with potential aeolian transport. *J. Mar. Sci. Eng.* **2019**, *7*, 357. [CrossRef]

16. Yokobori, M.; Kuriyama, Y.; Shimozono, T.; Tajima, Y. Numerical simulation of volume change on the backshore induced by cross-shore aeolian sediment transport. *J. Mar. Sci. Eng.* **2020**, *8*, 438. [CrossRef]

17. Sarre, R.D. Aeolian sand drift from the intertidal zone on a temperate beach: potential and actual rates. *Earth Surf. Process. Landf.* **1989**, *14*, 247–258. [CrossRef]

18. Delgado-Fernandez, I. Meso-scale modelling of aeolian sediment input to coastal dunes. *Geomorphology* **2011**, *130*, 230–243. [CrossRef]

19. De Winter, W.; Donker, J.; Sterk, G.; Van Beem, J.; Ruessink, G. Regional versus local wind speed and direction at a narrow beach with a high and steep foredune. *PLoS ONE* **2020**, *15*, e0226983. [CrossRef]

20. Lynch, K.; Jackson, D.W.T.; Cooper, J.A.G. Aeolian fetch distance and secondary airflow effects: The influence of micro-scale variables on meso-scale foredune development. *Earth Surf. Process. Landf.* **2008**, *33*, 991–1005. [CrossRef]

21. Rotnicka, J.; Dłużeski, M. A method to derive long-term coastal wind data from distant weather station to improve aeolian sand transport rate prediction. *Aeolian Res.* **2019**, *38*, 24–38. [CrossRef]

22. Hsu, S.A. Models for estimating offshore winds from onshore meteorological measurements. *Bound. Layer Meteorol.* **1981**, *20*, 341–351. [CrossRef]

23. Hsu, S.A. Correction of land-based wind data for offshore applications: A further evaluation. *J. Phys. Oceanogr.* **1986**, *16*, 390–394. [CrossRef]

24. Lynch, K.; Jackson, D.W.T.; Cooper, J.A.G. Foredune accretion under offshore winds. *Geomorphology* **2009**, *105*, 139–146. [CrossRef]

25. Bauer, B.O.; Davidson-Arnott, R.G.D.; Walker, I.J.; Hesp, P.A.; Ollerhead, J. Wind direction and complex sediment transport response across a beach-dune system. *Earth Surf. Process. Landf.* **2012**, *37*, 1661–1677. [CrossRef]

26. Walker, I.J.; Davidson-Arnott, R.G.D.; Bauer, B.O.; Hesp, P.A.; Delgado-Fernandez, I.; Ollerhead, J.; Smyth, T.A.G. Scale-dependent perspectives on the geomorphology and evolution of beach-dune systems. *Earth Sci. Rev.* **2017**, *171*, 220–253. [CrossRef]

27. Namikas, S.L.; Sherman, D.J. A review of the effcts of surface moisture content on aeolian sand transport. In *Desert Aeolian Processes*; Tchakerian, V.P., Ed.; Chapman and Hall: London, UK, 1995; pp. 269–293.

28. Schmutz, P.P.; Namikas, S.L. Measurement and modeling of the spatiotemporal dynamics of beach surface moisture content. *Aeolian Res.* **2018**, *34*, 35–48. [CrossRef]

29. Brakenhoff, L.B.; Smit, Y.; Donker, J.J.A.; Ruessink, G. Tide-induced variability in beach surface moisture: Observations and modelling. *Earth Surf. Process. Landf.* **2019**, *44*, 317–330. [CrossRef]

30. Delgado-Fernandez, I. A review of the application of the fetch effect to modelling sand supply to coastal foredunes. *Aeolian Res.* **2010**, *2*, 61–70. [CrossRef]

31. Van der Wal, D. Effects of fetch and surface texture on aeolian sand transport on two nourished beaches. *J. Arid Environ.* **1998**, *39*, 533–547. [CrossRef]

32. Hoonhout, B.; De Vries, S. Field measurements on spatial variations in aeolian sediment availability at the Sand Motor mega nourishment. *Aeolian Res.* **2017**, *24*, 93–104. [CrossRef]

33. Nickling, W.G.; Ecclestone, M. The effects of soluble salts on the threshold shear velocity of fine sand. *Sedimentology* **1981**, *28*, 505–510. [CrossRef]

34. Arens, S.M. Transport rates and volume changes in a coastal foredune on a Dutch Wadden island. *J. Coast. Conserv.* **1997**, *3*, 49–56. [CrossRef]

35. Ruz, M.H.; Meur-Ferec, C. Influence of high water levels on aeolian sand transport: upper beach/dune evolution on a macrotidal coast, Wissant Bay, northern France. *Geomorphology* **2004**, *60*, 73–87. [CrossRef]

36. Hage, P.; Ruessink, G.; Donker, J. Using Argus video monitoring to determine limiting factors of aeolian sand transport on a narrow beach. *J. Mar. Sci. Eng.* **2018**, *6*, 138. [CrossRef]

37. Turner, I.L.; Coates, B.P.; Acworth, R.I. The effects of tides and waves on water-table elevations in coastal zones. *Hydrogeol. J.* **1996**, *4*, 51–69. [CrossRef]

38. De Vries, S.; Van Thiel de Vries, J.S.M.; Van Rijn, L.C.; Arens, S.M.; Ranasinghe, R. Aeolian sediment transport in supply limited situations. *Aeolian Res.* **2014**, *12*, 75–85. [CrossRef]

39. Hage, P.; Ruessink, G.; Van Aartrijk, Z.; Donker, J. Using video monitoring to test a fetch-based aeolian sand transport Model. *J. Mar. Sci. Eng.* **2020**, *8*, 110. [CrossRef]

40. Cohn, N.; Hoonhout, B.M.; Goldstein, E.B.; De Vries, S.; Moore, L.J.; Durán Vinent, O.; Ruggiero, P. Exploring marine and aeolian controls on coastal foredune growth using a coupled numerical model. *J. Mar. Sci. Eng.* **2019**, *7*, 13. [CrossRef]

41. Wijnberg, K.M. Environmental controls on the decadal morphologic behaviour of the Holland coast. *Mar. Geol.* **2002**, *189*, 227–247. [CrossRef]

42. Aagaard, T.; Kroon, A.; Andersen, S.; Sorensen, R.M.; Quartel, S.; Vinther, N. Intertidal beach change during storm conditions; Egmond, The Netherlands. *Mar. Geol.* **2005**, *218*, 65–80. [CrossRef]

43. Stolk, A. *Zandsysteem Kust, Een Morfologische Karakterisering*; Geopro report 1989.02; Rijksuniversiteit Utrecht, Vakgroep Fysische Geografie: Utrecht, The Netherlands, 1989; 97p. (In Dutch)

44. Quartel, S.; Ruessink, B.G.; Kroon, A. Daily to seasonal cross-shore behaviour of quasi-persistent intertidal beach morphology. *Earth Surf. Process. Landf.* **2007**, *32*, 1293–1307. [CrossRef]

45. Schwarz, C.; Böhm, C.; Donker, J.; Ruessink, G. Observations of wind and sand transport across a vegetated foredune slope. *J. Geophys. Res. Earth Surf.* under review.

46. Ruessink, G.; Schwarz, C.S.; Price, T.D.; Donker, J.J.A. A multi-year data set of beach-foredune topography and environmental forcing conditions at Egmond aan Zee, The Netherlands. *Data* **2019**, *4*, 73. [CrossRef]

47. Guisado-Pintado, E.; Jackson, D.W.T. Multi-scale variability of storm Ophelia 2017: The importance of synchronised environmental variables in coastal impact. *Sci. Total Environ.* **2018**, *630*, 287–301. [CrossRef] [PubMed]

48. De Bruin, H.A.R.; Lablans, W.N. Reference crop evapotranspiration determined with a modified Makkink equation. *Hydrol. Process.* **1998**, *12*, 1053–1062. [CrossRef]

49. Wilson, G.W.; Fredlund, D.G.; Barbour, S.L. The effect of soil suction on evaporative fluxes from soil surfaces. *Can. Geotech. J.* **1997**, *34*, 145–155. [CrossRef]

50. Aydin, M.; Yang, S.L.; Kurt, N.; Yano, T. Test of a simple model for estimating evaporation from bare soils in different environments. *Ecol. Model.* **2005**, *182*, 91–105. [CrossRef]

51. Smit, Y.; Donker, J.J.A.; Ruessink, G. Spatiotemporal surface moisture variations on a barred beach and their relationship with groundwater fluctuations. *Hydrology* **2019**, *6*, 8. [CrossRef]
52. Davidson-Arnott, R. *An Introduction to Coastal Processes and Geomorphology*; Cambridge University Press: Cambridge, UK, 2010; p. 442.
53. Cohn, N.; Ruggiero, P.; De Vries, S.; Kaminsky, G.M. New insights on coastal foredune growth: The relative contributions of marine and aeolian Processes. *Geophys. Res. Lett.* **2018**, *45*, 4965–4973. [CrossRef]
54. Stockdon, H.F.; Holman, R.A.; Howd, P.A.; Sallenger, A.H., Jr. Empirical parameterization of setup, swash and runup. *Coast. Eng.* **2006**, *53*, 573–588. [CrossRef]
55. Kang, H.Y.; Nielsen, P. Water table dynamics in coastal areas. In Proceedings of the 25th International Conference on Coastal Engineering, Orlando, FL, USA, 2–6 September 1996; pp. 4601–4612.
56. Van Genuchten, M.T. A closed-form equation for predicting the hydraulic conductivity of unsaturated soils. *Soil Sci. Soc. Am. J.* **1980**, *44*, 892–898. [CrossRef]
57. Schmutz, P.P.; Namikas, S.L. Measurement and modeling of moisture content above an oscillating water table: implications for beach surface moisture dynamics. *Earth Surf. Process. Landf.* **2013**, *38*, 1317–1325. [CrossRef]
58. Hsu, S.A. Wind stress criteria in eolian sand transport. *J. Geophys. Res.* **1971**, *76*, 8684–8686. [CrossRef]
59. Shao, Y.P.; Lu, H. A simple expression for wind erosion threshold friction velocity. *J. Geophys. Res.* **2000**, *105*, 22437–22443. [CrossRef]
60. Iversen, J.D.; Rasmussen, K.R. The effect of surface slope on saltation threshold. *Sedimentology* **1994**, *41*, 721–728. [CrossRef]
61. Kok, J.F.; Partell, E.J.R.; Michaels, T.I.; Karam, D.B. The physics of wind-blown sand and dust. *Rep. Prog. Phys.* **2012**, *75*, 72. [CrossRef]
62. De Vries, S.; Southgate, H.N.; Kanning, W.; Ranasinghe, R. Dune behavior and aeolian transport on decadal timescales. *Coast. Eng.* **2012**, *67*, 41–53. [CrossRef]
63. Ruessink, B.G.; Arens, S.M.; Kuipers, M.; Donker, J.J.A. Coastal dune dynamics in response to excavated foredune notches. *Aeolian Res.* **2018**, *31*, 3–17. [CrossRef]
64. Smit, Y. Surface Moisture Dynamics on a Narrow Beach. Ph.D Thesis, Utrecht University, Utrecht, The Netherlands, 2019.
65. Hage, P.M.; Ruessink, B.G.; Donker, J.J.A. Determining sand strip characteristics using Argus video monitoring. *Aeolian Res.* **2018**, *33*, 1–11. [CrossRef]
66. Hsu, S.A. Computing eolian sand transport from routine weather data. In Proceedings of the 14th Conference on Coastal Engineering, Copenhagen, Denmark, 24–28 June 1974; pp. 1619–1626.
67. Pollock, L.W.; Hummon, W.D. Cyclic changes in interstitial water content, atmospheric exposure, and temperature in a marine beach. *Limnol. Oceanogr.* **1971**, *16*, 522–535. [CrossRef]
68. Yang, Y.; Davidson-Arnott, R.G.D. Rapid measurement of surface moisture content on a beach. *J. Coast. Res.* **2005**, *21*, 447–452. [CrossRef]
69. Bauer, B.O.; Davidson-Arnott, R.G.D.; Hesp, P.A.; Namikas, S.L.; Ollerhead, J.; Walker, I. Aeolian sediment transport on a beach: Surface moisture, wind fetch, and mean transport. *Geomorphology* **2009**, *105*, 106–116. [CrossRef]
70. Namikas, S.L.; Edwards, B.L.; Bitton, M.C.A.; Booth, J.L.; Zhu, Y. Temporal and spatial variabilities in the surface moisture content of a fine-grained beach. *Geomorphology* **2010**, *114*, 303–310. [CrossRef]
71. Atherton, R.J.; Baird, A.J.; Wiggs, G.F.S. Inter-tidal dynamics of surface moisture content on a meso-tidal beach. *J. Coast. Res.* **2001**, *17*, 482–489.
72. Turner, I.L. Water table outcropping on macro-tidal beaches: A simulation model. *Mar. Geol.* **1993**, *115*, 227–238. [CrossRef]
73. Oblinger, A.; Anthony, E.J. Surface moisture variation on a multibarred macrotidal beach: implications for aeolian sand transport. *J. Coast. Res.* **2008**, *24*, 1194–1199. [CrossRef]
74. Houser, C. Synchronization of transport and supply in beach-dune interaction. *Prog. Phys. Geogr.* **2009**, *33*, 733–746. [CrossRef]
75. Duarte-Campos, L.; Wijnberg, K.M.; Hulscher, S.J.M.H. Estimating annual onshore aeolian sand supply from the intertidal beach using an aggregated-scale transport formula. *J. Mar. Sci. Eng.* **2018**, *6*, 127. [CrossRef]
76. Arens, S.M. Rates of aeolian transport on a beach in a temperate humid climate. *Geomorphology* **1996**, *17*, 3–18. [CrossRef]

77. Jackson, N.L.; Nordstrom, K.F. Aeolian transport of sediment on a beach during and after rainfall, Wildwoord, NJ, USA. *Geomorphology* **1998**, *22*, 151–157. [CrossRef]
78. van Dijk, P.M.; Stroosnijder, L.; de Lima, J.L.M.P. The influence of rainfall on transport of beach sand by wind. *Earth Surf. Process. Landf.* **1996**, *21*, 341–352. [CrossRef]
79. Sherman, D.J.; Jackson, D.W.T.; Namikas, S.L.; Wang, J. Wind-blown sand on beaches: An evaluation of models. *Geomorphology* **1998**, *22*, 113–133. [CrossRef]
80. Van Rijn, L.C.; Strypsteen, G. A fully predictive model for aeolian sand transport. *Coast. Eng.* **2020**, *156*, 19. [CrossRef]

Journal of
Marine Science and Engineering

Article

Comprehensive Natural Environment and Landscape Signs in Coastal Settlement Hazard Assessment: Case of East Taiwan between the Coastal Mountain and the Pacific Ocean

Shyang-Woei Lin [1,*], Chia-Feng Yen [2], Chih-Hsin Chang [3], Li-Jin Wang [4] and Hung-Ju Shih [3]

[1] Department of Natural Resources and Environmental Studies, National Dong-Hwa University, Hualien 97401, Taiwan
[2] Department of Public Health, Buddhist Tzu-Chi University, Hualien 97004, Taiwan; mapleyeng@gmail.com
[3] Slopeland & Hydrology Division, National Science and Technology Center for Disaster Reduction, New Taipei City an Téaváin 231, Taiwan; chang.c.h@ncdr.nat.gov.tw (C.-H.C.); shihruby713@ncdr.nat.gov.tw (H.-J.S.)
[4] National Dong-Hwa University College of Environmental Studies, Disasters Prevent Center, National Dong-Hwa University, Hualien 97401, Taiwan; foryou2607@gmail.com
* Correspondence: shine@gms.ndhu.edu.tw; Tel.: +886-3-890-5184

Received: 17 May 2020; Accepted: 25 June 2020; Published: 28 June 2020

Abstract: In East Taiwan, coastal settlements are scattered and narrowly confined between the Coastal Mountain and the Pacific Ocean. These settlements are currently at risk as there is no room for retreat. Therefore, it is essential to conduct a comprehensive and continuous hazard assessment in these coastal residential areas. In order to avoid biased towards the natural environment, the factors that cannot easily be built within the geographic information system (GIS) database are distinguished by Unmanned Aerial Vehicle (UAV) to conduct a vulnerability assessment of threats to coastal zones. The method: we used the east coast of Taiwan as an example, through GIS and statistical analysis in land-use status, vulnerable population groups and UAV landscape signs of indicators of erosion and accumulation. Through the main output of an intuition scatter map, the erosion landscape susceptibility, economical land-use exposure, and special population groups' ratio allowed for the easy comparison of the vulnerability, risk level and resilience between different coastal settlements. These diverse observation aspects of risk assessment results can provide prevention and control strategies that meet the different needs of coastal risk management in restricting and strengthening the land-use development of communities.

Keywords: coastal hazard; vulnerability assessment; unmanned aerial vehicle; landscape

1. Introduction

Two percent of the world's surface is 10 meters above sea level. However, more than 10% of the world's population lives in these islands, low-lying coastal areas and deltaic coastal areas [1,2]. In Taiwan, the sea-level change rate is much higher than the global average [3]. Waves with heights of three to five meters are common when the northeast monsoon blows along the coast in winter, and the typhoons in summer often cause huge waves to reach seven meters. These particularly affect the East Coast of Taiwan, which is comprised of the Coastal Mountain and is bordered by the Pacific Ocean. The East Coast contains scattered coastline settlements for more than 100 kilometers. Under the threat of global warming, it continues to be exposed to risks such as sea-level rise, coastal zone erosion, storm surges, and extreme rainfall [4].

One effective way to prevent coastal environmental disasters is to design coastal receding or disaster buffer zones by limiting land use and population growth in disaster-prone areas. Taking Central and South America as examples, under the conditions of different aspects of economy, ecology and disaster, development has been restricted in coastal areas that are between 10 to 200 meters. [5].

Taiwan's "Coastal Management Plan" restricts development in coastal areas by distinguishing its coast into first-class coastal protected areas in which development is forbidden, and second-class protected areas which can be developed conditionally after review [6]. There is no first-level coastal protection zone on the eastern coast of Taiwan, and three coastal sections the longest area designated as a second-class protected areas, totaling 109 kilometers and comprising 31.4% of the eastern coastline, are classified within the second-level protection zone [7]. It is clear these descriptions that the coastal erosion problem is common for all the settlements discussed in this article.

However, the natural environment has a very limited plain area, and the management plan does not offer an effective solution for the prevention of environmental disasters. Restricting land use and development in offshore land settlements requires more detailed landscape and land-use surveys and education to local populations on the importance of respecting the environment to limit disasters.

The hazard assessment-incorporated basic data of the coastal area settlements was obtained from an environmental survey using remote sensing and GIS to refine the resolution of coastal assessments to the nearby coastal settlements. Considering four aspects of the coastal settlements, namely the risk level of erosion of the landscape, the environmental vulnerability of the land use, peoples' exposure at the time of the disaster, and the ratio of old and young people in the settlement population, the coastal settlements were visualized in an intuition scatter bubble chart using the assessment results. This allowed us to provide a solution for disaster avoidance without detailed information on the coastal settlements.

The research area is over 100 km long and 1 km wide, encompassing narrow coastal settlements of eastern Taiwan located near the Coastal Mountain and the Pacific Ocean. The following coastal settlements were considered in this study: Shoufeng Township's Fudekeng, Yanliao, Ganzishu, and Shuilian, Fengbin Township's Jiqi, Guian, Xinfeng, Libi, Xinshe, Xinzhuang, Dongxing, Yongfeng, Fengbin, Lide, Lifu, Shiti, Shitiping, Gangkou, Dagangkou, Jingpu, and Sanfu, totaling 21 settlements.

2. The Evaluation of the Vulnerability of Coastal Areas Review

The Reviewing Office of the United Nations Disaster Relief Coordinator (UNDRC), The UN Office for Disaster Risk Reduction (UNDRR), and Intergovernmental Panel on Climate Change (IPCC) research reports on environmental disasters in coastal areas show that the concept of vulnerability is constantly adjusted over time. For example, vulnerability analyzed by the UNDRC considers the risk of static socio-economic factors such as population, architecture, economy, and infrastructural elements [8]. UNDRR emphasizes the susceptibility of these natural, social, economic and environmental factors on the impact of harm [9]. In addition to the factors that increase susceptibility, the IPCC also proposes the concepts of coping, adaptability, and resilience [10].

In 1979, a static concept of abstraction and vulnerability of potential natural hazards that occurred in a specific period and region. In 2004, this concept was developed to include the dynamic measurement of the sensitivity of the economic or human life dynamics to increased loss exposure and risk level, more emphasis on adaptability and resilience after disasters. Until now, emphasizing disaster reduction and sustainable development is to respond to the aforementioned process of static, dynamic, and resilient reconstruction in order to achieve the goal of "more disaster resistance". For example, UNDRR, "Sendai Framework for Disaster Risk Reduction 2015–2030" passed by the meeting of the third World Disaster Reduction Conference was held in Sendai, Miyagi Prefecture, Japan, March 2015 [11].

More recent studies focused on the evaluation of the vulnerability of coastal areas make use of the Coastal Vulnerability Index (CVI), which is calculated using a combination of six parameters, namely terrain, coastline change, coast slope, relative sea-level rise, wave height, and tidal range, and is widely used in many countries [12–17]. The advantage of using the CVI and applying GIS and remote sensing data in spatial analysis is the rapid understanding of the natural environment of coastal areas. However, as can be seen from the literature, the inadequacy of using only one point of view and two indices were estimated—not only the single aspect of CVI indicators, but also the Social Vulnerability Index (SVI) in the assessment of the vulnerability of the coastal settlement environment [18]. It can also prove that this article, using comprehensive multiple aspects indicators, socio-economic environmental factors and landscape is a better way of coastal settlement hazard assessment.

3. Research Method

It can be learned from the literature review mentioned above that the assessment of coastal erosion in settlements requires different considerations. The first indicator of this study uses UAV to look for signs of erosion or the deposition of landscape, which is the first comprehensive risk level aspect. Calculating the possible economic loss when settlements are affected by disasters in coastal areas from the GIS database of land-use survey is the second environmental vulnerability aspect. The population exposures of settlements and the ratio of old and young people in the settlement of vulnerable groups from statistics is the third aspect.

3.1. Indicators of Erosion and Accumulation

The factors that contribute to the dynamic change of the coastal environment are not only from the physical environment of nature, such as earthquakes, tsunamis, storm surges, waves, sea-level rise, and tectonic changes, but also from various human activities, including newborn land formation along the coast due to large breakwaters concrete or sand extraction, coastal sand drifting blocked by large coastal structures, reduction of river sediment transport by reservoirs and sand dams, and the over-pumping of groundwater by coastal aquaculture fisheries [6].

While such coastal disasters cannot be interpreted by a single factor, they are often attributable to the combined effects of land utilization by humans, and the interactions between coastal protection and the natural environment, maintaining a delicate dynamic balance in these sensitive and fragile coastal areas.

In France, the Regional Coastal Hazards Observatory (OR2C) uses drones for the coastal land monitoring of the erosion of the Atlantic coast [19]. Drone surveys thus provide valuable information for assessing the factors of erosion and accumulation of coastlines cannot easily been built within the GIS database. This study therefore made use of a drone aerial photography survey of landscape signs in coastal settlements to assess coastal vulnerability, as shown in Figure 1.

(a)

(b)

Figure 1. *Cont.*

Figure 1. Examples of landscape signs and drone aerial photography survey in coastal settlements. (**a**) large area of planted windbreaks built along the coast; (**b**) coastal landforms due to orographic uplift; (**c**) multilayer concrete or offshore breakwaters; (**d**) erosion scarps along the shoreline; (**e**) several bank revetments that seem to be newly built

Furthermore, disaster-inducing risks to coastal settlements were evaluated using the indices of erosion and deposition signs to respond to the more direct impact brought by these complex environments. Each symptom factor is given a point. The more the settlement's coastal pattern or landscape meets the description of the erosion index, the higher the score of the erosion sign. The score for signs of erosion minus points for deposition signs give a total score, indicating that this settlement is subject to a slight risk of erosion. Conversely, if the settlement coast has more items that match the deposition landscape, the deposition score will be higher.

For each coastal settlement within the scope of the study, the aforementioned methods were used to obtain an evaluation score of coastal settlement erosion signs in Table 1. The checklist of cause can be adjusted according to the actual situation depending on the survey area and special conditions or different countries.

Table 1. Drone aerial photographs of signs corresponding to ground survey photos.

The Coast is Under Erosion if the Following Phenomena are Observed on the Landscape	
Physical environment	Human activities
visible dune erosion scarps along the shoreline	multilayer concrete hollow squares or offshore breakwaters
without evident deposition, while the area is in the vicinity of an estuary	several bank revetments that seem to be newly built
few unstable new and low plants in the seaward dune plant line ecology, and stable fruticose in the first row	several debris barriers, river sand mining, or groundwater abstraction in the upstream
seascape, sea notch, wave-cut platform, sea cave, marine arch, and sea stacks in the coastal landform	undermining formed due to protruding bank revetments or coastal cape in the upper littoral current, and the sand imbalance areas formed thereby
sea waves pouring over slope breakwater in several places behind the windbreak	settlement farming is mainly rice, which uses more water, and irrigation water is mostly from groundwater.
coastline has receded significantly in the past 10 years, identified from Google Earth satellite images	interviewees or residents think that coastline is severely eroded
The coast is under deposition if the following phenomena are observed	
Physical environment	Human activities
a large area of sand exposed at low tide	storm line or seaside garbage line that is far away from the coastline, with a large area of wet and dry sand beaches inside waves
significant sediment accumulation at the estuary	old bank revetments that are nearly covered by sea sand
a large area of new and low plants at the first line of coastal vegetation; coastal terrace, wave-cut platform	sea reclamation in several places along the coast, and a large area of planted windbreaks or several bamboo fences built along the coast
coastal landforms due to orographic uplift	spit-type sand accumulation effect due to bank revetments in the lower littoral current
lagoon, pools, and other elevated coastal or deposition landforms offshore	settlement farming was originally rice, and it was converted to dry crops such as tomatoes or cereals with little water.
coastline has shifted forward in the past 10 years, identified from Google Earth satellite images	interviewees or residents think that the coastline is heavily deposition.
all signs of the erosion and deposition landscapes of coastal settlements as a result, of totaling the scores points □ 1–2 points slightly □ 3–4 points moderate □ 5–6 points severe □ over 7 points very serious	

3.2. Evaluation of the Extent of Exposure to Coastal Hazards from the Settlement Land Use

Taking into account the rapid changes in the socio-economic environment and actual needs in Taiwan, the Ministry of the Interior promotes a case-by-case investigation and mapping of land use every 10 years, starting in 1990. This study uses the results of the most recent (2015) land-use survey. The existence of very fine scales and well-categorized land-use survey data for the country allow for a detailed calculation of the disaster potential hazard areas and possible resultant economic losses.

Countries without survey data can use satellite imagery to perform similar calculations. Therefore, for coastal settlements only, according to the extent of the vulnerability of coastal erosion, three major land-use categories can be identified: construction, transportation and open space, as displayed in Table 2.

Table 2. Settlement area showing different land-use patterns within the coastal land area.

Categories	Open Space	Construction	Transportation	Total (m^2)
Settlement Areas [1]	35,272	117,972	43,570	196,814
Settlement Percentage	18%	60%	22%	100%

[1] Areas is been calculated from statistics of land-use investigation of Taiwan [20].

The aforementioned land-use survey is the result of a detailed land-use classification system. For example, the transportation category includes the public facilities in the settlements, the construction category is highly exposed to disasters, and open space represents low-level development in the face of disasters. The area percentage of these categories in the settlement area represents the economic loss of the settlement after the disaster in Figure 2. This method can be appropriately adjusted according to the resolution of existing land-use data. Vulnerability refers to a system's exposure to damage or capacity for loss for a given hazard [21]. This study used the percentage of construction area in the coastal settlement as an indication of the vulnerability of settlements.

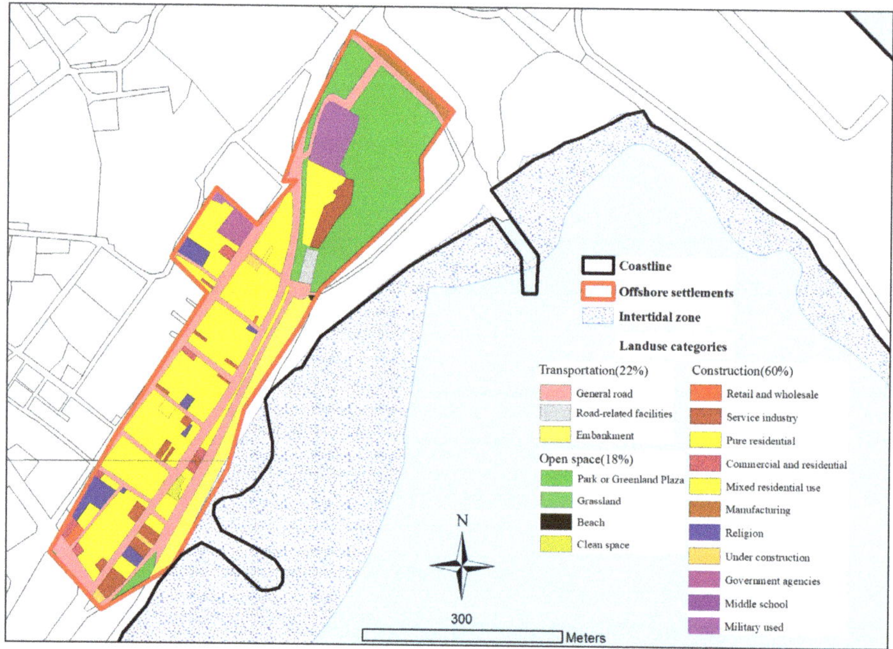

Figure 2. Land-use category area distribution within the settlement area.

3.3. Estimation of the Number of People in the Settlement Area

The settlement area that may be affected by coastal erosion is usually not the smallest unit of statistics. There is no direct population or special age group statistics. Therefore, it must be calculated from GIS estimations of address matching or rooftop number.

Many countries, including Taiwan, have point coordinates for each household's address, therefore, after overlaying the area of the settlement polygon on the map in GIS, the total number of households in the settlement can be calculated. According to the statistical data, the average population of each household in the village in the smallest statistical unit provides an accurate estimation of the population of the settlement. In the same way, an estimation of the population under 14 years old and over 65 years old, who will struggle to a greater extent to recover after disasters, can be obtained.

In countries where only satellite imagery is available, the actual number of roofs on each street block within the settlement can be counted. Using the ratio with the smallest statistical unit, the actual population in each settlement can also be deduced, as shown in Figure 3.

(a)   (b)

Figure 3. Population estimation by means of rooftop counting. (**a**) Orthophoto or satellite image; (**b**) Rooftop digitizing.

4. Results

In summary, the coastal disaster risk assessment of this research incorporates four parameters determined from remote sensing and GIS, as shown in Table 3. The first factor is the use of drone and ground landscape surveys to comprehensively determine the degree of erosion and deposition of coastal settlements. The second factor is the calculation of high-risk areas, such as construction or economic areas, and the loss as a percentage of the total settlement land-use area using environmental survey data. In addition to the effects of possible exposure on properties, the third factor involves the population of the settlement, thereby addressing the exposure of human life. The fourth factor involved understanding the population ratio of vulnerable residents, the old and young, when the risk of coastal hazards between different settlements is high.

Table 3. Assessment of the different settlements in the study area.

Settlement.	Signs from Landscape [1]			Land Use of Constructions [2]		Population Number [3]	Ratio of Vulnerable Residents [3]		
	Erosion	Deposition	Erosion Minus Depositions	Areas (m^2)	Percentage within Settlement	Total (People)	Under 15 young	Older than 65 elder	Total Percentage Within Settlement
Fudekeng	4	1	3	12,298.62	27%	156	21	4	17%
Yanliao	4	1	3	82,193.3	36%	208	29	4	17%
Ganzishu	2	4	2	17,442.16	27%	115	16	2	17%
Shuilian	5	1	4	97,701.58	33%	617	152	5	33%
Jiqi	4	2	2	25,315.11	44%	111	17	4	21%
Guian	3	1	2	14,567.71	30%	102	15	3	20%
Xinfeng	4	1	3	3,656.65	46%	29	5	4	24%
Libi	3	1	2	4,549.13	39%	40	7	3	23%
Xinshe	6	1	5	26,057.13	36%	212	39	6	25%
Xinzhuang	4	1	3	1,853.1	30%	21	4	4	24%
Dongxing	3	2	1	24,218.57	42%	158	29	3	24%
Yongfeng	4	1	3	28,832.33	35%	133	26	4	27%
Fengbin	0	1	1	153,001.76	62%	879	175	0	27%
Lide	3	0	3	12,298.62	27%	116	23	3	28%
Lifu	4	1	3	4,822.75	47%	25	5	4	28%
Shiti	3	2	1	16,181.66	7%	181	40	3	33%
Shitiping	3	2	1	17,451.35	33%	125	28	3	33%
Gangkou	2	1	1	48,406.46	53%	395	88	2	33%
Dagangkou	1	2	1	41,779.52	41%	109	24	1	32%
Jingpu	2	4	2	96,195.73	44%	590	130	2	28%
Sanfu	1	3	2	17,720.4	53%	98	22	1	29%

[1] each coastal settlements' signs of the erosion and deposition landscapes is from the NSTCDR [22]; [2] areas are from land-use investigation of Taiwan [20], [3] population number and vulnerable residents are been calculated from official statistics until 2019 Jan.

In Figure 4, the Y-axis is used as the percentage of construction land use that has the highest vulnerability to coastal erosion in the settlement; the X-axis is the percentage of the vulnerable young and elderly population in the settlement. The population of the settlements is represented by the size of the symbol, which allows for a visual comparison of the threats of erosion to different population sizes in the same coastal environment. The color represents the degree of coastal erosion and deposition in the landscape survey. Red represents erosion and green represents deposition, with deeper shades representing higher degrees of erosion/deposition. This system allows for a good risk assessment of the different settlements.

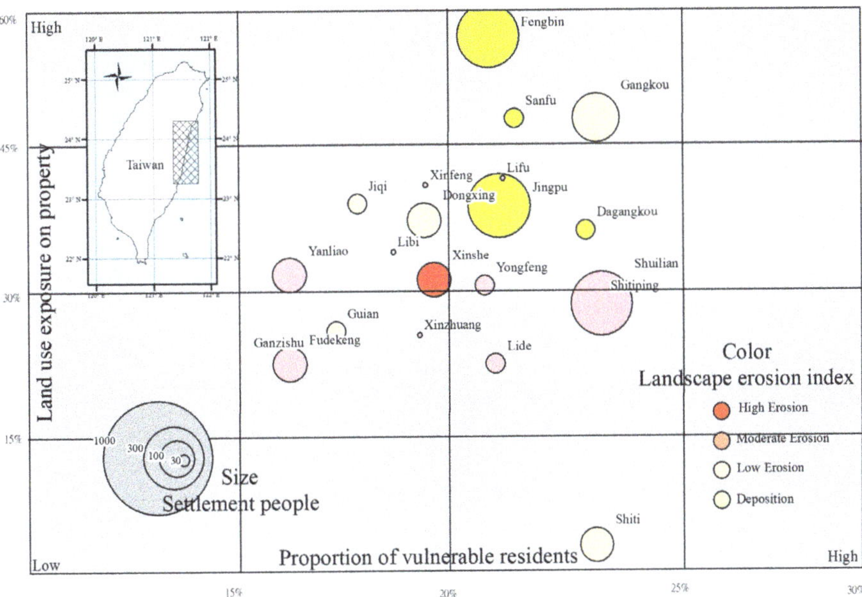

Figure 4. Coastal settlement vulnerability assessment on the eastern side of Taiwan

Taking Xinshe settlement as an example, the red circle representing the settlement indicates that the landscape is encountering severe coastal erosion. In addition, the construction land-use ratio and the vulnerable population percentage are both median compared with other settlements in the same environment, therefore Xinshe falls in the center of the bubble chart. In contrast, Fengbin settlement has a relatively large population, so the circle diameter is large; the coast is in a state of deposition, and the land is mostly highly utilized for construction and economic purposes, therefore this settlement falls in the upper part on the bubble chart. As the population of the Shuilian settlement population is large, the symbol is also large; the proportion of young and elderly people is high, hence it occurs on the right-hand side of the chart; the coastal settlement is moderately eroded, therefore the circle is light orange-red in color.

5. Discussion and Conclusions

The accessibility of the Taiwan eastern coastal environment is not as good as further inland, and the difficulty of a comprehensive environmental investigation is relatively high. Therefore, existing investigation methods must be used to determine the coastal impact on the population of the local settlements.

The use of drones, which are a relatively cheap and popular survey tools, are highly effective for use in research involving remote sensing image surveys. In just 10 minutes from take-off to landing, drones capture hundreds of high-resolution orthographic or side shots. The coastal landscape formed as a result of numerous erosional and depositional events occurring in the area. Landscape survey methods are used to explore these high volumes of images, and interviews with residents as well as field surveys can assist in assessing the risk level of erosion for each nearshore settlement. Simply treating each erosion and deposition factor equally may not reflect the decisive influence on the coastal settlement. However, intuition is the biggest advantage of this method. The judgment criteria's checklist can also be adjusted in different coastal environments per local environmental conditions. This method of quantifying the landscape checklist through drone photography and on-site surveys serve as an important reference for assessing the risk level threat of coastal erosion facing coastal settlements.

The second way to understand the coastal settlement is to assess the data from land-use surveys. This is a very suitable method in Taiwan where comprehensive environmental surveys exist for the island. The investigation of land use in Taiwan has been actively promoted since the establishment of the "National Land Information System Basic Environment Construction Plan" by the central government in the 1990s, so it has accurate and complete survey data with a high resolution and integrity. Therefore, it can be used as a calculation basis for the exposure of construction land use under the threat of coastal erosion as for the concept of vulnerability faced by coastal settlements. The categorization of land use into three categories, namely construction, transportation, and open space, allows for this method to be used in other countries where survey data do not exist and satellite images with lower resolutions are used. The results indicate that it is already possible to correctly and effectively judge the economic development of near-shore settlements. To determine the actual economic loss experienced by coastal settlements due to erosion, a more detailed calculation of economic loss using drone images to investigate is necessary.

Observing the degree of threat to settlements from coastal disasters using various assessment factors, it can be seen that different settlements are significantly affected, which means that the threats to coastal settlements should not be evaluated in a single aspect.

In the assessment of coastal disasters, in addition to the assessment of landscape and land-use economic property loss, this study also focused on the awareness of disasters in coastal settlements. Considering that resilience is one of the main focuses of modern environmental disaster research when disasters occur, residents need to rely on their disaster prevention, responses and recovery. Therefore, the age composition of the elderly and young should be determined. The population ratio of these vulnerable groups with a certain degree of correlation should be considered in the face of disaster resilience. However, survey data such as population or age composition can usually only be counted in the minimal administrative unit, the village, and cannot be directly applied in an environment where a village may include multiple coastal settlements.

The third way to understand coastal settlements from a comprehensive dataset is to assess census survey data. Using orthophoto images, we can directly calculate the number of households or roofs in the settlements, and then estimate the statistical data by determining the average number of people per household in the village. From the results of the study, it can be seen that this estimation is very close to the real numbers, and therefore, the method employed is an effective method.

The results of the study also indicate that this method of assessing the hazard of the coastal settlements is not only intuitive, effective and fast, but also meets the needs of assessing the exposure and resilience of vulnerable coastal settlements. The study area takes the eastern coast in Taiwan most susceptible to coastal erosion as an example. The results show that, from the observed indicators, in addition to the impact of the natural environment, under the influence of other human activities, the influence is no less than natural factors, and it also has a strong impact on coastal erosion. In addition to the CVI or GIS numerical database survey, coastal disaster assessment can be combined with aerial photography landscape methods.

Author Contributions: Conceptualization, S.-W.L.; Data curation, C.-F.Y.; Funding acquisition, C.-H.C.; Investigation, S.-W.L. and L.-J.W.; Methodology, S.-W.L.; Project administration, S.-W.L.; Resources, C.-H.C. and H.-J.S.; Validation, C.-H.C.; Writing–original draft, S.-W.L.; Writing–review & editing, C.-F.Y. All authors have read and agree to the published version of the manuscript.

Funding: This research was funded by "National Science and Technology Center for Disaster Reduction" in Taiwan, grant number NCDR-S-108018.

Acknowledgments: We would like to thank valuable comments from anonymous reviewers and all participants who made this study possible.

Conflicts of Interest: The authors declare no conflict of interest.

References

1. Helderop, E.; Grubesic, T.H. Social, geomorphic, and climatic factors driving US coastal city vulnerability to storm surge flooding. *Ocean Coast. Manag.* **2019**, *181*, 104902. [CrossRef]
2. McGranahan, G.; Balk, D.; Anderson, B. The rising tide: Assessing the risks of climate change and human settlements in low elevation coastal zones. *Environ. Urban.* **2007**, *19*, 17–37. [CrossRef]
3. Tseng, Y.H.; Breaker, L.C.; Chang, E.T.Y. Sea level variations in the regional seas around Taiwan. *J. Oceanogr.* **2010**, *66*, 27–39. [CrossRef]
4. IPCC. *Global Warming of 1.5 °C*; United Nations: New York, NY, USA, 2018; pp. 1–616.
5. Perez-Cayeiro, M.L.; Chica-Ruiz, J.A.; Garrido, M.A.; Bedoya, A.M. Revising the limits of the coastal area in the regulations of the iberoamerican region. Are they appropriate for risk management and adaptation to climate change? *Ocean Coast. Manag.* **2019**, *181*, 104912. [CrossRef]
6. Interior, M.O.T. *Integrated Coastal Management Plan*; National Science and Technology Center for Disaster Reduction: Taipei, Taiwan, 2017; pp. 1–195.
7. The 9th River Management Office, Water Resource Agency. *Hualien Coastal Protection Plan*; NCKU Hydraulic and Ocean Engineering R&D Foundation: Tainan, Taiwan, 2019.
8. UNDRO. *Nature Disasters and Vulnerability Analysis*; United Nations Disaster Relief Coordinator: New York, NY, USA, 1979; p. 8.
9. UNDRR. *Living with Risk: A Global Review of Disaster Reduction Initiatives*; United Nations Inter-Agency Secretariat of the International Strategy for Disaster Reduction: Geneva, Switzerland, 2004; p. 16.
10. IPCC. *Climate Change 2014: Impacts, Adaptation, and Vulnerability Part A: Global and Sectoral Aspects*; Press, C.U., Ed.; Intergovernmental Panel on Climate Change: New York, NY, USA, 2014; p. 5.
11. UNDRR. *Sendai Framework for Disaster Risk Reduction 2015–2030*; United Nations Special Representative of the Secretary-General for Disaster Risk Reduction: Geneva, Switzerland, 2015; p. 35.
12. Djouder, F.; Boutiba, M. Vulnerability assessment of coastal areas to sea level rise from the physical and socioeconomic parameters: Case of the Gulf Coast of Bejaia, Algeria. *Arab. J. Geosci.* **2017**, *10*, 299. [CrossRef]
13. Hereher, M.E. Coastal vulnerability assessment for Egypt's Mediterranean coast. *Geomat. Nat. Hazards Risk* **2015**, *6*, 342–355. [CrossRef]
14. Hoque, M.A.; Ahmed, N.; Pradhan, B.; Roy, S. Assessment of coastal vulnerability to multi-hazardous events using geospatial techniques along the eastern coast of Bangladesh. *Ocean Coast. Manag.* **2019**, *181*, 104898. [CrossRef]
15. Islam, M.A.; Mitra, D.; Dewan, A.; Akhter, S.H. Coastal multi-hazard vulnerability assessment along the Ganges deltaic coast of Bangladesh-A geospatial approach. *Ocean Coast. Manag.* **2016**, *127*, 1–15. [CrossRef]
16. Rajan, S.M.P.; Nellayaputhenpeedika, M.; Tiwari, S.P.; Vengadasalam, R. Mapping and analysis of the physical vulnerability of coastal Tamil Nadu. *Hum. Ecol. Risk Assess.* **2019**, 1–17. [CrossRef]
17. Karymbalis, E.; Chalkias, C.; Chalkias, G.; Grigoropoulou, E.; Manthos, G.; Ferentinou, M. Assessment of the sensitivity of the southern coast of the Gulf of Corinth (Peloponnese, Greece) to sea-level rise. *Cent. Eur. J. Geosci.* **2012**, *4*, 561–577. [CrossRef]
18. Tragaki, A.; Gallousi, C.; Karymbalis, E. Coastal hazard vulnerability assessment bsed on geomorphic, oceanographic and demographic parameters. *Land* **2018**, *7*, 56. [CrossRef]
19. Kerguillec, R.; Audere, M.; Baltzer, A.; Debaine, F.; Fattal, P.; Juigner, M.; Launeau, P.; Le Mauff, B.; Luquet, F.; Maanan, M.; et al. Monitoring and management of coastal hazards: Creation of a regional observatory of

coastal erosion and storm surges in the pays de la Loire region (Atlantic coast, France). *Ocean Coast. Manag.* **2019**, *181*, 104904. [CrossRef]

20. National Land Surveying and Mapping Center. Land Use Investigation of Taiwan. 2019. Available online: https://www.nlsc.gov.tw/LUI/Home/Content_Home.aspx (accessed on 1 January 2020).

21. Grubesic, T.H.; Matisziw, T.C. A typological framework for categorizing infrastructure vulnerability. *Geojournal* **2013**, *78*, 287–301. [CrossRef]

22. National Science and Techonolgy Center for Diaster Reduction (NCDR). *Coastal Settlement Environment Investigation and Disaster Risk Assessment Research*; National Dong Hwa University College of Environmental Studies Diaster Prevention Center: Taiwan R.O.C., 2019.

MDPI

St. Alban-Anlage 66

4052 Basel

Switzerland

Tel. +41 61 683 77 34

Fax +41 61 302 89 18

www.mdpi.com

Journal of Marine Science and Engineering Editorial Office

E-mail: jmse@mdpi.com

www.mdpi.com/journal/jmse

9 783036 504964

When
All
Means
All